I0059265

The Inside of
Outsourcing

The Inside *of* Outsourcing

A Pragmatic View From The Inside

John C. LaBella

Copyright © 2012 By John C. LaBella,

All rights reserved.

Published in the United States by LCI, an imprint of LCI Publishing Group. Madison, Wisconsin

ISBN 978-0-9855536-0-9

Printed in the United States of America MMXII.

Illustrations and cover graphics artwork provided created by Kurt Schoenfielder, Throttle 5 Design. Throttle5.com

Copy Editing Services provided by Angela Romig.

Content Editing provided by Sheri Supena.

Final Manuscript Review by Carl LaBella.

About the Author

John C. LaBella is President of LaBella Consulting, Inc. He is dedicated to helping firms work through the many issues encountered as they attempt to optimize their Information Technology (IT) functions through outsourcing. John has traveled the globe working with partners in the U.S., Mexico, Brazil, Eastern Europe, Western Europe and India.

Prior to managing IT organizations, John led major manufacturing and logistics initiatives at Kraft Foods, Inc. John taught Operations Management at the *University of Wisconsin, Madison - School of Business* and at *Edgewood College* in the MBA program. John was invited to lecture at *L'Université de Bordeaux* in Bordeaux, France and also at the Massachusetts Institute of Technology, Cambridge, MA.

In his book, ***The* Inside *of* Outsourcing,** John shares his insights and offers antidotes from the past 15 years working in various IT roles. He has held positions ranging from Director of IT, Kraft Foods, Inc. to Vice President of Information Technology, Gap, Inc. During this time, John developed a great interest in, and respect for, the outsourcing of IT services.

***The* Inside *of* Outsourcing** culminates John's experiences into a practical "must read" for executives searching for ways to improve their IT organizations. The book is also a practitioner's guide designed to take the reader through the varied activities leading to a successful outsourcing experience.

The process of outsourcing Information Technology Services to offshore partners can be daunting, especially the first time. Through ***The* Inside *of* Outsourcing** John offers the reader a pragmatic look at the inner workings of outsourcing projects, how to do them and how to stay out of trouble.

Acknowledgments:

I would be remiss if I didn't take a moment to recognize some of the team leaders that I have been fortunate to work with over the years. These people have each, in their own way, provided the inspiration and input needed to write *The* Inside *of* Outsourcing. I would like to extend a warm "thank you" to the following individuals for their efforts: John Gawin, Pauline Cheung, Srikanth Krishna, Susana Chan, Arun Raghavapudi, Dinesh Bajaj, Sheri Supena, Ed Symington, John Chiang, Gaurav Agarwal. And last, but not least, fellow consultant, Michael Tey, with whom I spent endless hours discussing the finer points of outsourcing while inside the finer restaurants in San Francisco.

Over the past few years they have all offered ideas, support and a unique opportunity to help me understand *the inside of outsourcing...*

Dedication:

This book is dedicated to my family for allowing me to travel the world during my corporate career and to spend all those endless mornings at the local coffee shop writing this book.

Amy, Jessi, Beth and Chris – you guys continue to make me proud.

A special heart felt thank you to my loving wife and lifelong cheerleader, Patti. Without her spirit, inspiration and energy, I would not be the person I am today.

Professional Perspective

"This book is an amazing collection of many different scenarios and examples explained in a simple and effective way to understand how to transition into structured outsourcing that best benefits organizations.

Too many times we are working either at client or supplier organizations and mostly see the point of view from only one side. Great effort to present a balanced view as seen from client and supplier viewpoints to ensure a successful and long-lasting relationship is presented.

This book can be a handbook for anyone who is involved with managing outsourcing deals".

Srikanth Krishna

Implementation Manager

Bangaluru, India

The Inside *of* Outsourcing

Table of Contents

The *Inside* of *Outsourcing*

Forward: *"The Inside of Outsourcing"*

The intended audience for this book are executives looking to drive efficiencies in their IT organizations. *The Inside of Outsourcing* will explain how to develop and implement well crafted strategies to optimize their IT labor model. The book examines strategic planning, implementation and project evaluation processes to keep the organization out of trouble and moving forward.

The Inside of Outsourcing is also a comprehensive step-by-step guide giving those leading the outsourcing initiative an improved chance for success by providing a pragmatic path to follow from initial concept development to implementation. The concepts and techniques in this book have been field-tested a number of times. The book is a culmination of many years of successful outsourcing.

Over the past 15 years, we have observed dramatic growth in the outsourcing of Information Technology (IT) services. India based IT services firms such as Infosys (INFY), WIPRO (WIT) and Tata Consulting Services (TCS) have seen their revenues explode while providing a broad array of service offerings in the global IT outsourcing market. The annual revenue for INFY in 1999 was $100 Million. Considering INFY was founded in 1983, growing to $100 Million in a little over 15 years is very respectable. However, by 2008 their revenues grew to an excess of $4 Billion (USD). Now that is nothing short of awe-inspiring!!

Every IT professional recalls the panic caused by the Year 2000 (Y2K) bug. IT departments around the globe were scrambling for resources to "find & fix" suspect code as the Year 2000 rapidly approached. The aforementioned Indian firms (and others) were able to provide low-cost resources with requisite technical proficiency to address this effort. More importantly, these firms were process oriented which proved to be the difference in how efficiently they found and repaired legacy software code. Much of their work was done "offshore" in India, utilizing the capabilities afforded by the expanding Internet. Y2K and the invention of the web browser were the primary catalysts for the success of these firms, and hence, the growth in offshore outsourcing.

The Inside of Outsourcing focuses on the outsourcing of IT services from an insider's perspective. It examines how firms can approach the topic and design rational, effective strategies that will result in the successful implementation of outsourcing solutions. The book is segmented into four areas of focus: Building the Business Case; Organizational Readiness; Supplier Selection; and Implementation & Governance.

Areas of Focus: "*The* Inside *of* Outsourcing"

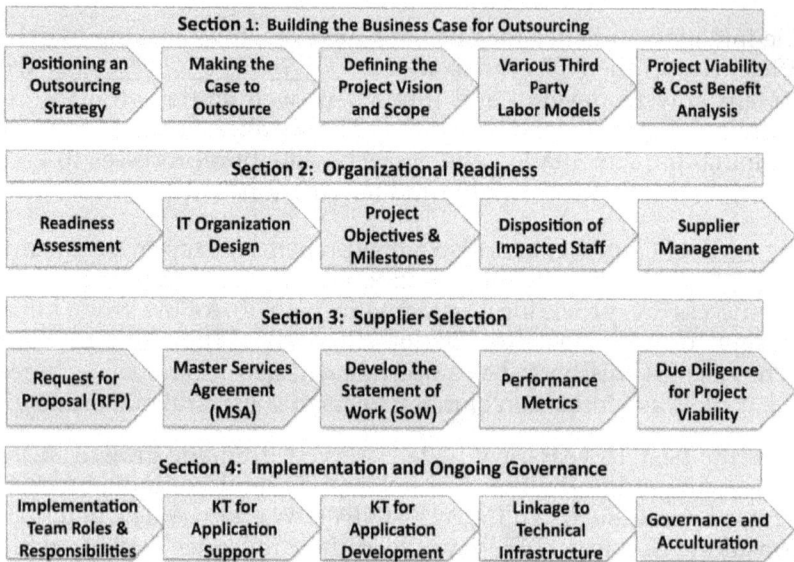

Section 1: Building the Business Case for Outsourcing				
Positioning an Outsourcing Strategy	Making the Case to Outsource	Defining the Project Vision and Scope	Various Third Party Labor Models	Project Viability & Cost Benefit Analysis

Section 2: Organizational Readiness				
Readiness Assessment	IT Organization Design	Project Objectives & Milestones	Disposition of Impacted Staff	Supplier Management

Section 3: Supplier Selection				
Request for Proposal (RFP)	Master Services Agreement (MSA)	Develop the Statement of Work (SoW)	Performance Metrics	Due Diligence for Project Viability

Section 4: Implementation and Ongoing Governance				
Implementation Team Roles & Responsibilities	KT for Application Support	KT for Application Development	Linkage to Technical Infrastructure	Governance and Acculturation

This book will not address outsourcing the manufacturing of physical products nor will we delve into Business Process Outsourcing. Companies that have been successful moving manufacturing offshore follow a similar pattern to what we will be discussing in this book. They look to outsource noncore activities that can be done more effectively by someone else. Although the move to offshore manufacturing is well documented, firms will not hesitate to manufacture in the U.S. when market conditions and cost profiles dictate. For example, CISCO outsources the manufacture of a large percentage of the products it sells. However, they opened a very large (2500 employees) fiber optics facility in the state of New Hampshire a few years ago because it was beneficial for them to do so.

The primary focus of *The Inside of Outsourcing* will be outsourcing and "off-shoring" Application Support and Application

Development activities. The IT organization needs labor to deliver the systems and services they provide. IT labor can either be direct employees, temporary staffing used for a short period of time, and/or outsourcing work to a third party partner. As with manufacturing, a primary reason for outsourcing IT services to offshore partners is to help contain growing operating costs within the organization.

The Inside of Outsourcing is structured and written as a "How To" book to help firms successfully prepare to outsource. The book will benefit both the provider and receiver of IT services by improving the success rate of outsourcing initiatives.

During my career in the corporate world I held many management roles in IT ranging from Director of Application Development to the Vice President of Application Support Services. I was fortunate to have participated in and led major outsourcing initiatives at two Fortune 500 companies. Although they operate in very different industries, the strategies and tactics used by these firms to execute successful outsourcing are very similar. The book offers the reader strategic themes, analysis and actions needed to deliver their objectives. The observations contained in it are applicable and easily adaptable to any company regardless of size or business sector. *The Inside of Outsourcing* shares many of the "pitfalls" I experienced along the way as well. Hopefully the reader can learn from them.

Without the tireless efforts and exemplary leadership of the many people that I have had the pleasure of working with over the years, the road would have been a much more difficult one to travel.

And now, please enjoy *The Inside of Outsourcing* . . .

The *Inside* of *Outsourcing*

Executive Insights

What to watch for when considering outsourcing

➢ Pre-ignition check-list:

- Define the problem you would like outsourcing to solve. Every problem may not find its solution in outsourcing.

- Identify and prepare to dedicate individuals to manage the evaluation. Use the same folks to manage whatever solution they suggest – it never hurts to have skin in the game.

- Understand how anticipated future costs compare to current costs. This requires a deep dive into the current labor model and a thorough cash-flow analysis.

- Secure funding approval for money needed for one-time implementation costs such as travel, Knowledge Transfer, severance, data lines between onshore and offshore, etc.

➢ Risk assessment:

- Outsourcing benefits are well publicized and include cost reduction, resource flexibility, improved quality, and more.

- Define the problem - don't start outsourcing until the *problem* that outsourcing will solve has been defined.

- Outsourcing needs to fit into the IT strategic plan. It must be well planned rather than a knee-jerk reaction.

- Understand the risks of using Staff Augmentation over a long period of time vs migrating the organization to a Managed Services Model.

- Outsourcing is more than capturing attractive offshore rates. It will have an impact on work dynamics for IT.

- Dedicate structured teams to create and execute the strategy. This cannot be done well if added to existing workloads.

- Prepare to manage relationship changes that occur between IT and the business processes that rely on technology.

- Make sure there is a plan in place for retaining and managing knowledge that may be lost through outsourcing.

- Use high quality partners in acceptable geographies.

- Prepare to manage the impact that outsourcing will have on existing staff. Many will either be let go or redeployed. It is hard to achieve savings if nobody is taken off the org chart.

- With a Managed Services Model, the organization needs to learn how to manage *outcomes* instead of *activities*.

- The Governance Team must change their work style from *doing* to *reviewing*, as they will be managing the partner using Service Level Agreements.

- Outsourcing to offshore partners means learning how to work across multiple time zones.

- Take steps to minimize the impact of the supplier gaining too much leverage.

- Build language into the agreement to assure that there can be an amicable split if/when the relationship sours.

- Guard against retained staff becoming disenfranchised.

➤ Financial drivers:

- Secure incremental funding required for travel, Knowledge Transfer, severance, consulting and other misc. costs before launching the project.

- It is difficult and costly to undo an area that has been outsourced.

- Manage risk by starting in a small area first rather than going big initially.

- Spend the time to fully understand the cost structure of internal staff as well as the cost of the supplier's resources.

- Become familiar with the concept of blended cost.

➤ Organizational readiness:

- A Call Management Center is essential for outsourcing the Application Support function. This should be implemented well ahead of outsourcing support.

- Check for other major initiatives that may conflict with the outsourcing project. On the other hand, those other initiatives may be able to utilize impacted staff that will be freed up by outsourcing.

- It is critical to keep the team focused with clear objectives.

- Centralizing the support staff into a stand-alone team clarifies focus and drives superior results when outsourcing support.

- Determine how to manage the reduction of internal staff.

- The perspective changes from managing lines of code to managing outcomes. Managers need to relearn how to write Statements of Work that reflect outcomes.

- Managing suppliers in general may be optimized by a more formal supplier management function.

➢ Finding the right supplier:

- The Request for Proposal (RFP) process needs to be structure to provide a fair and thorough assessment of available suppliers.

- Be careful not to "seed" the answer. Allow the team to build their assessment and recommend a supplier.

- Develop a top-to-top relationship with the chosen supplier to build rapport that can lead to innovation and growth.

➢ Developing the contract and performance metrics:

- The Legal department needs to drive this process and use the Working Team to help understand the desired outcomes of the agreement.

- It is best if the Master Services Agreement (MSA) is in place before allowing a supplier to participate in the RFP. The MSA must be in place before the supplier is allowed to negotiate a contract.

- The baseline data for performance metrics should be able to reflect the capabilities of the current process.

➢ Implementation:

- From the date the contract is signed, the time required to plan and implement an outsourcing effort could easily span 16 – 24 weeks depending on the size and complexity of the project.

- Don't rush or try to shorten the time needed for knowledge transfer. Once the new resources are trained, the current staff will be gone along with their knowledge.

- Throughout the transition there will be checkpoints (also called "reverse knowledge transfer) to test how well the new resources have learned their roles.

➤ Governing the relationship:

- Governance is an ongoing process to assure that long-term partnership between the firm and the supplier is nurtured and maintained in a positive manner.

- With outsourcing projects, the number of Governance Team resources needed ranges anywhere from 10-20% of the total number of resources supporting the outsourced function.

- Make sure the Governance Team is staffed with a good mix of employees with solid technical skills and business knowledge.

- Ideally, the Working Team becomes the Governance Team once the implementation is complete.

Section 1:

Building the Business Case

For Effective Outsourcing

The first hurdle to effective outsourcing is making sure there is a valid business reason to outsource in the first place. In Section 1, we will take emotion out of the equation by focusing on fact and data to build the business case. Topics discussed will include:

➤ Positioning outsourcing as a component of a comprehensive labor strategy for Information Technology (IT) services

➤ Cost drivers and resource constraints in the IT organization

➤ Defining the project by first understanding the specific problem to be solved

➤ Exploring various third party labor models such as consulting, staff augmentation, and managed services

➤ Using actual data to develop a cash-flow model that will validate the business case

Areas of Focus: *"The* Inside *of* Outsourcing"

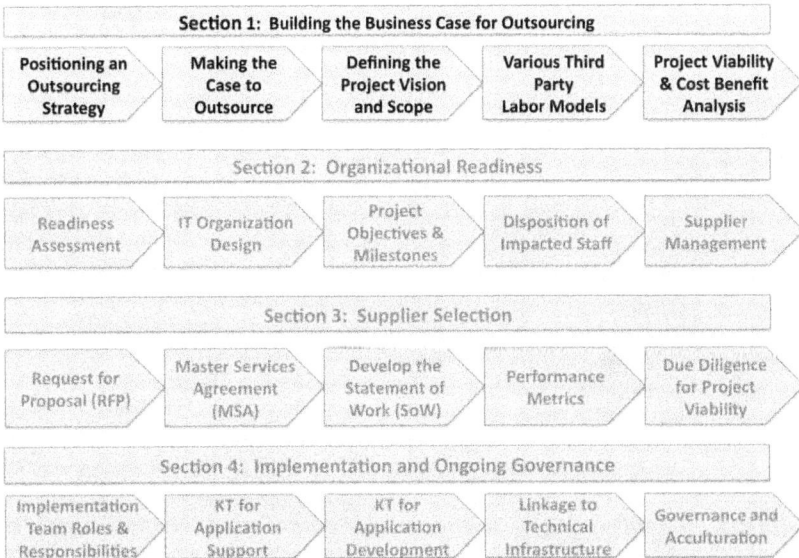

Section 1: Building the Business Case for Outsourcing				
Positioning an Outsourcing Strategy	Making the Case to Outsource	Defining the Project Vision and Scope	Various Third Party Labor Models	Project Viability & Cost Benefit Analysis

Section 2: Organizational Readiness				
Readiness Assessment	IT Organization Design	Project Objectives & Milestones	Disposition of Impacted Staff	Supplier Management

Section 3: Supplier Selection				
Request for Proposal (RFP)	Master Services Agreement (MSA)	Develop the Statement of Work (SoW)	Performance Metrics	Due Diligence for Project Viability

Section 4: Implementation and Ongoing Governance				
Implementation Team Roles & Responsibilities	KT for Application Support	KT for Application Development	Linkage to Technical Infrastructure	Governance and Acculturation

The *Inside* of *Outsourcing*

Chapter 1:

Positioning an Effective Outsourcing Strategy

Outsourcing is not a new idea – it was the catalyst that propelled civilization from hunter-gathers to our current global based economy.

At one point, a caveman decided that it was easier to grow his own corn than it was to walk around looking for it. As his efficiencies in corn growing flourished, he soon found he had more corn that he could use. At the same time, the caveman in the next cave found that he could grow wheat in great abundance. The cavemen exchanged their surplus with each other and outsourcing was born.

Look around. How many firms today generate their own electricity or drill their own water supply? Certainly there are some that do when it is economically beneficial to do so, but most don't. Why? They rely on the efficiencies provided by local utilities for power and water. These utilities focus on a specialized "processes" resulting in a depth of expertise and efficiencies to optimize the delivery of their product.

We don't think of outsourcing basic commodity services as unusual. But many firms don't go beyond this to analyze what other functions or services lay beyond their core services and the efficiencies they can achieve by outsourcing these to specialized, efficient providers.

Admittedly, outsourcing power and water is a "no brainer." Let's dig a little deeper into other important activities and processes that are commonly outsourced and try to understand why. Think about the number of times you have worked late and there, like clockwork, appears Wanda. Wanda is the person that enters your personal workspace to empty the trash, vacuum the carpet, dust the furniture, arrange your family pictures, etc. From time to time you may have had conversations with Wanda. Over time, she learns just how you like your workspace cleaned, when to dust, what to touch and what not to disturb, etc. Unlike power and water, one can relate to janitorial services (or more specifically the janitor) on a personal level. Yet, in many cases, janitorial services are outsourced for the same reason that power and water are outsourced. These are commodity processes that can be provided at a lower cost by a firm that specializes in providing them. The same is true of other non-core processes such as: cafeteria services, lawn care and building maintenance.

What I find interesting is that in many cases, we may not even be aware which processes are outsourced – and we don't really care. Who provides these services is not important. What *is* important is whether or not the service can be provided at a favorable price with acceptable quality when compared to providing them in-house.

To be sure, there is someone in the firm responsible for managing the relationship with its service providers. After all, someone needs to represent the company and "govern" the relationship to assure the consistent delivery of service and quality. When things go awry, corrective action needs to be taken to put things back on track.

OK – so there is no angst about outsourcing mundane processes. But what about those processes that are more important to the business?

Payroll processing is an activity that is near and dear to my heart. When it fails, I am directly impacted along with others who rely on their paycheck. However, payroll processing is frequently outsourced. There are a number of firms, such as ADP, that have been providing this service for years. Most employees are unaware of this and as long as their paycheck is delivered as planned, they don't really care. Payroll processing is very important and the key to why it can be outsourced is that it is a "non-core" or "commodity" process. Other activities that have been successfully outsourced include: Engineering, Employee Benefits, Recruiting, Transportation Planning, Logistics, Legal, Compensation and IT Services to name a few. When activities are outsourced, make sure someone in the firm is responsible for governing the relationship with the provider.

The skeptics may say, "I agree that outsourcing commodity work is common place but what about the jobs lost? What will become of the displaced employees?" To understand the answer to this question, we need to examine the concept of "specialization" and its impact on the broader marketplace.

The business world has been migrating away from a vertically integrated model. Vertical integration means that a company manages most of its operations internally. Replacing these vertically integrated organizations is a web of specialized partners delivering essential, but non-core functionality so the firm can focus on its core processes. To better understand outsourcing we will examine the opposite end of the spectrum and discuss the ultimate "in-sourcing" – the company town.

During the mid to late 1800's, the "company town" concept grew across the United States. It was the antithesis of outsourcing in that everything in the town was owned and maintained entirely by the company. Many services were provided to employees in hopes of keeping them well housed, happy and loyal – this was also called the "The Utopian Worker Community." A good example of this is Pullman (a neighborhood on the south side of Chicago, Illinois). At its peak, the Pullman Company employed about 6,000 people manufacturing luxury passenger train cars. The company provided housing, a library, entertainment, churches and retail markets for the employees and their dependents. Employees rented their homes from the company and were required to live in the Pullman owned "company town" even though cheaper housing could be found in nearby communities. George Pullman's dream of a utopian worker community ended with violent labor riots and a bitter strike in 1894. Four years later in 1898, the Supreme Court of Illinois ruled that the Pullman Co. had to dissolve their ownership in the town and allow its residents to purchase their homes from the company. At the peak of the company town era there were nearly 2,500 of them scattered across the country housing 3% of the U.S. population.

Early in my career, I worked for a large food processing company. This company was vertically integrated and managed nearly every aspect of their supply chain. They grew the grain, developed feed blends, raised the breeding stock, managed the feed-lots, slaughtered the animals, separated meat from the carcass, formulated product specifications, blended spices, developed marketing strategies, created their own packaging, manufactured the finished product and eventually distributed it to grocery stores so it could be sold in the market place. Over time, each step along the supply chain came under scrutiny as the firm looked for ways to improve their overall efficiency. Processes were examined searching for ways to provide a lower cost alternative at the same or better quality. Today, the company relies on a conglomeration of third party business partners instead of the vertical "supply chain."

As with the company town example, firms that had once been vertically integrated are outsourcing more and more of their commodity processes to third party partners. To be sure, the problem of governing the relationships with the partners is important and will be addressed later in the book. However, the productivity contributions derived from utilizing specialized business partners is simply too great to ignore. The term "specialized," refers to a firm that focuses on a piece or subset of the overarching process. By focusing on a small domain, new firms are created resulting in higher efficiency, effectiveness and quality. These "new firms" become partners providing better value than what was otherwise attainable. Outsourcing to specialized partners has allowed many firms to remain competitive and deliver a higher value to the market place. Economics 101 taught us that as the cost goes down demand goes up requiring more production. Our economy works best when productivity improvements lower the cost of goods for everyone. Productivity is simply "more output from the same or less input." The new firms create jobs and hire employees that may have been displaced by aforementioned outsourcing initiatives. In their new jobs, these employees are able to buy products (maybe those of the food company) and soon everyone lives happily ever after. And so the cycle goes…

Enter the skeptic again! "I get it. Cost management and productivity are needed for firms to compete and survive. But what about those high paying jobs (such as IT jobs) that leave the U.S. and go offshore?" It is true that a small percentage of IT jobs in the food company example did in fact get outsourced to an offshore partner. However, their products along with the products of many other manufacturing companies in this country are sold "offshore". As discretionary incomes rise in countries we outsource jobs to, so does their ability to buy more and more of the products we sell.

We need to realize that we live and work in a global economy. To deny that is burying one's head in the sand. When jobs move offshore, those employees are able to purchase things made in the U.S. This in turn results in job growth in the U.S. in order to produce what the rest of the world is buying from us.

According to the U.S. Census Bureau, the value of all exports from the U.S. to India was $8 Billion (USD) in 2005 and that number more than doubled to $21.6 Billion by 2011. U.S. exports to China grew from $41 Billion in 2005 to nearly $103 Billion in 2011. Then U.S. President, Barack Obama, set a five-year goal of doubling exports by 2014. As U.S. exports grow, so do domestic jobs. People had to be hired to produce the incremental $124 Billion dollars worth of exported goods going to India and China. The increase in the export of manufactured goods from 2010 to 2011 added 404,000 jobs in the U.S. Currency valuations impact exports and imports. When the dollar value is low, it drives exports to countries with higher comparative currency values.

Survival in the global economy cannot be legislated or restricted by governmental interference. To survive, firms need to provide a high quality product at a price that creates value for the customer regardless of where in the world that product is produced or sold.

The argument that the U.S. should bring all those outsourced IT jobs back to this country is flawed by the shear magnitude of the numbers. Yes, the unemployment rate in the U.S. in 2011 was over 9% and grew steadily since 2009. However, the unemployment rate from 2000 to 2008 averaged 5.2% and during that time the demand for quality IT resources outstripped the supply in the U.S. Since we had a difficult time finding and hiring candidates in this country, outsourcing IT work to offshore firms grew steadily. It has been estimated that $200 - $300 billion per year is spent on outsourced IT services and represents about two to three million IT jobs. Offshore outsourcing will not end tomorrow. If it did, there is no way to find the resources in this country to fill all the open positions. When demand is greater than supply, competition increases resource costs. If total IT spending is fixed, at a higher cost, fewer resources can be hired than we had before. This results in fewer productivity projects and a reduction in growth.

The long-term viability of every private company relies on profit growth. Without it, the firm soon withers and dies and all jobs are lost. Profit growth relies on cost containment, increasing sales volume, and consistent quality of the products and services being delivered. In today's world, outsourcing is often part of a firm's cost reduction strategy. Successful outsourcing starts by segmenting internal processes into core and noncore (commodity) and identifying those that can be performed more efficiently by others. Once again, let me emphasize that noncore activities are important (if they were not, they shouldn't be done in the first place) but often can be performed more effectively by an outside partner. Core activities on the other hand are of strategic importance to the firm and directly impact the product or services the firm provides. Successfully outsourcing the "right" processes will help improve profitability as well as the quality of the product and/or services provided.

In his book The World is Flat, Pulitzer Prize winning author Thomas L. Friedman states: "Several technological and political forces have converged, and that has produced a global, Web-enabled playing field that allows for multiple forms of collaboration without regard to geography or distance or soon, even language." These technological breakthroughs – primarily the internet, web browsers and work-flow tools have made the ability to outsource IT services to remote offshore locations a reality. Caution: Outsourcing to offshore partners should not be a de facto decision. It must be evaluated and planned properly. A strategy that simply sends all IT services to India can be a cataclysmic mistake. Selecting what to outsource (and knowing what NOT to outsource) is a critical first step in developing the business case for outsourcing IT services.

Outsourcing, especially a large initiative, will significantly change how work is done within and around the IT function. It will also impact those business processes that are tightly linked to IT tools. Outsourcing doesn't happen on its own, and to assure acceptance and success, it must be well managed. In order to maximize results, it is recommended that a three-tier team structure be created. Table 1-1 outlines a three-tier team structure that has been very effective in the outsourcing engagements I have participated in.

The very first requirement to assure outsourcing success is alignment across the executive stakeholders. Whether or not to outsource IT services needs to be an integral part of the CIO's strategic plan for the IT organization. As such, it must be consistent with the overall objectives of the firm as well. To assure alignment, an Executive Steering Team (EST) should be formed that includes the CIO, CFO, and other senior executives from functions that the outsourcing initiatives will impact (i.e. Manufacturing, Distribution, HR, Finance, etc).

Table 1-1 Outsourcing Teams

Teams	Roles & Responsibilities
Executive Steering Team	- Create the vision and define the problem to be solved - Develop Guiding Principals - Decide which geographies to look at - Recommend contract guidelines and spending criteria - Review/approve WT recommendations
Working Team	- Project definition and scope - Select the appropriate sourcing model - Conduct preliminary cost/benefit analysis - Review project with EST for approval - Finalize objectives and timelines - Manage impacted staff - Communicate project across the organization - Implement Knowledge Management - Manage RFP, Supplier Selection, Contract and Governance - Manage Knowledge Transfer-
Process Team	- Perform day-to-day work activities - Validate Service Level Agreements (SLA) data - Manage Knowledge Transition - Identify issues with supplier personnel - Oversee application or technical upgrades

Executive Steering Team:

The Executive Steering Team is responsible for creating the high level vision for IT outsourcing. Often, the desire is to utilize third party partners to reduce cost, improve quality and generate innovative ideas to build business value in the marketplace. To achieve this requires a commitment from the Executive level to build a cohesive "top-to-top" relationship with designated partners.

Before a problem can be solved, it first must be defined. The Executive Steering Team needs to articulate what that problem is. The problem may be the difficulty the firm is facing finding specific technical skills in local markets. The problem may be that rapid technology change is making it difficult to keep employees trained and proficient on leading edge tools. The problem may be that the firm is frustrated with the cost and time required to deliver IT projects with traditional internal staff.

Guiding principals need to be created to shape the direction that new outsourcing initiatives will take. For example, they need to decide how much of the total IT budget should be spent on outsourcing. Guidance is needed on the right number of partners to engage – is one to few, is 20 too many? Does the firm wish to identify geographies that are "off limits" because of socio-economic concerns such as currency stability, political unrest, and the state of governmental relations with the U.S. The country or region selected must be able to produce enough high quality graduates to supply the projected demand for them.

Contract guidelines will help to assure that the legal interests of the firm are protected and all Terms and Conditions align with legal department directives. The Executive Steering Team should define the maximum length of the contract and financial exposure the firm will allow itself to be bound to. They should also identify those applications and processes that will not be included in the scope of the contract.

The Executive Steering Team should provide financial and spending guidelines for working teams to follow. Examples include the Internal Rate of Return hurdle for cash flow modeling and approvals required for various spending levels in future Statements of Work. The most important need is to establish how any incremental expense associated with outsourcing will be funded. There will be an investment required to transfer knowledge from incumbent resources to the new partner. Make sure the funding vehicle is figured out before the Working Team spends the next few months framing the project.

The pacing and content of updates from subordinate teams needs to be established as well. These updates should be done on a regular basis to assure top-down alignment and consistent direction. Initially, monthly updates are needed to keep the project moving along quickly. Once the project is complete, quarterly updates should be fine.

Working Team:

The Working Team, as the name implies, is the team doing most of the work-of-the-work associated with outsourcing. The Working Team oversees day-to-day efforts required to effectively deliver an outsourcing project. This team should be led by a senior IT executive along with team members representing IT, Finance, Human Resources, Procurement and Legal.

The Working Team needs to define the scope and vision at the project level. This should mesh with the over-arching vision provided by the Executive Steering Team. In essence, they need to define the specific work activities that will be outsourced.

There are a number of different sourcing models to consider. The Working Team will evaluate those models and review options such as Managed Services or Staff Augmentation with potential partners. They also need to determine if they will use a fixed bid or time and materials approach to pay for supplier services.

The project objectives and high level timelines for completing them need to be drafted. They should also create a preliminary assessment of how many current resources will be impacted by the outsourcing decision along with an estimated cost for anticipated severance.

As they work through the facts/data associated with the defined problem, the Working Team should create a cost/benefit analysis for the project. The finance team member should manage this, as it is the first checkpoint to discuss with the Executive Steering Team to decide if it still makes sense to continue the project.

If the Executive Steering Team gives their approval to go forward, the Working Team should create a communication plan in order to make sure the broader organization is informed. Keeping stakeholders and IT staff informed throughout the entire initiative is key to eliminating detours in the future.

The loss of internal knowledge is the single biggest risk to outsourcing. The Working Team will need to define a methodology for Knowledge Management and decide how both tacit and explicit knowledge will be captured, retained and shared with those who need it in the future. Tacit knowledge is that which is learned by doing and is rarely in a format that can be recorded. This is also called "tribal knowledge" and is usually in someone's head. An example of tacit knowledge is how to hit a baseball – one learns this by doing it. Explicit knowledge on the other hand is specific information that can be codified or recorded in a table or database. An example of tacit data would be Federal tax rates found on one of the IRS look-up tables.

Supplier Selection and contract preparation (details in Section 3) will also be managed by the Working Team. They will conduct supplier interviews and manage the Request for Proposal (RFP) process. Once the supplier is identified, they will assist the legal team with elements needed for the Master Services Agreement (MSA) as well as definitions for the Operating Level Agreements, Service Level Agreements and Statements of Work (SoW).

The Working Team is responsible for Implementation and Knowledge Transfer for the project. They will manage the transition and resolve conflicts as they occur. Of major importance is finding and assigning individuals to specific implementation roles in order to support the timelines for the Knowledge Transfer phase. This is key to creating a solid foundation leading to project completion or steady state.

Once the project is complete the Working Team morphs into the Governance Team. Objective for the Governance Team include: transition objectives; cost targets; scope changes; dispute resolution; and a management escalation process. The Governance Team will also monitor the data supporting Service Level Agreements, Key Performance Indicators and Operating Level Agreements.

Process Team:

The Process Team will form just prior to the start of implementation. Naming the individuals on the PT should not happen until the project has been approved. These resources will actually do the day-to-day work associated with the project. This team is the foundation for effective outsourcing. It should be staffed with retained client staff to manage the transition and provide ongoing governance. The supplier's resources will also be part of the Process Team. Activities performed by the Process Team result in a "delivered" project.

The Process Team manages the Knowledge Transition process to assure supplier resources are adequately trained and perform in a competent manner. To do this they must assure that each training session has an existing Subject Matter Expert (SME) assigned to facilitate Knowledge Transfer. A process called reverse KT assures that each new resource has been properly trained will test the new resources. The Process Team assures that training materials are built, updated and used regularly by the supplier.

Incident Management is still the primary responsibility of the Process Team during transition since many of them will come from the existing support organization. In this dual role, they need to assure that all incidents are resolved as quickly as possible during transition and escalate as required to upper management. During the transition they may identify opportunities to revise scripting for the Call Management Center to assure that tickets are routed efficiently.

Since the Process Team is working side-by-side with the new resources they will be in the best position to raise any personnel issues pertaining to poor performance, language skills, inadequate technical knowledge, etc. to upper management.

Process Teams also have requisite knowledge to review application and/or technical upgrades, identify the impact to the environment and take appropriate actions to assure successful installation.

The world is a large place and determining the specific geographies and countries to outsource to is an important decision. I am not going to tell you where or where not to outsource. That is purely your decision and should be made after a thorough analysis of basic selection criteria such as:

Political stability:

Eliminate countries where the government is unstable and on the brink of collapse. Even though the labor rates may be low, the cost you will pay to rapidly exit that country if/when things get really bad will be very high. Remember, once work is outsourced it is critical for the client company to visit the offshore facility often. A good rule is to limit outsourcing to countries that you would feel safe traveling in.

Language skills:

One of the more intriguing aspects of India as an outsourcing location is the fact that it was once part of the British Empire. Because of this, the English language has been taught in Indian schools for years. It is rare to find an educated Indian citizen that is not proficient in formal English. Although the accent of some Indians may be a bit difficult to decipher by our Western ears, I have found that after working with them for a while I have adapted to it. Locations such as China, Brazil, the Philippines, and others have not had the opportunities to evolve the use of English as widely as it is in India. However, they are quickly trying to improve this situation. I suggest that each Statement of Work (SoW) you generate with an offshore outsourcing firm contains criteria regarding proficiency in the English language. Include the ability for you to reject resources that are difficult to understand.

Language skills are especially critical for Call Center outsourcing. Remember, even with strong IT skills, if the person in that Call Center in Pune, India is not able to be understood, they provide negative value to your customer and ultimately to you. On the other hand, language skills become less important if the offshore staff is providing technical support or developing software code. In these cases, they are more likely to be talking directly with partner staff working in this country.

Infrastructure:

The growth of the Indian economy (8.5% annually as of this writing) has quickly outstripped the ability of their infrastructure to keep up. The Indian government has committed to doubling their investment in infrastructure for roads, utilities, and mass transit to one Trillion (USD) over the five years from 2012 - 2017. Major 4 lane highways are rare in India. They tend to build a few miles at a time and then divert traffic onto non-surfaced dirt roads. Severe traffic congestion in the cities has been a major hassle for anyone who lives and/or visits India. On one of my visits, two people on a camel (yes, expect to see lots of animals on the roads – cows, monkeys, oxen and even camels) were actually moving faster than the taxi I was in.

When 70-80% of your IT outsource partner is located in some remote corner of the world, you can't afford to have your voice and data communications cut due to technical difficulties. With development partners there is also a large a mount of data moving back and forth. Chose countries with reliable electric power or a supplier that can generate their own to provide reliable connection to the Internet. Remote access capability is something your team really needs to nail down. It is what allows that person 10,000 miles away to operate as if they were in the room with you. Voice and data transmission needs to be well planned and executed as a foundation element in your outsourcing plans...it doesn't just "happen." A high-caliber outsourcing partner will be able to provide a lot of help as you develop your infrastructure solution. They have been doing it for years and many have redundant networks complete with Disaster Recovery plans – ask them to explain it to you during the Request for Proposal phase of the project. Interestingly, most of the infrastructure issues I faced were caused by problems on the U.S. end of the connection. We will discuss various strategies to mitigate this risk later in the book.

Educational facilities and IT graduate production:

The education system in India and China produce approximately 1.0 Million engineering graduates per year (450,000 and 550,000 respectively). This sounds impressive however; only 25-30% actually score high enough to be considered as potential IT candidates. At the crux of the issue is a lack of qualified instructors. As recently as two to three years ago, the Indian education system produced only fifty PhD candidates in IT per year. This resulted in many students being under-qualified to pass proficiency exams. It is difficult for these countries to keep up with the projected demand for IT services and one of the reasons why wage inflation for IT skills is currently projected to be 10% per year.

➢ The larger Indian IT outsourcing firms have taken matters into their own hands. They have invested heavily in facilities to provide extended education to capable graduates providing the training needed to become productive. I visited an entire campus in India built by one of the suppliers. They are able to train and house 15,000 employee-students at a time. The facility is complete with state-of-the-art classrooms, lecture halls, dormitories, swimming pools, movie theaters, and restaurants. Students live on campus and enjoy all the comforts of a five star resort while learning new technologies and skills. The education of these employees continues on an ongoing basis even after they are deployed to client projects. Expect to have a mix of seasoned resources as well as a few "freshers" working on your projects (freshers are less experienced resources). Make sure the experience level of these resources is referenced in the contract and SoW's.

Wage and price inflation:

The good news for India and China, as of this writing, is that their economies are growing in the 8-9% range. The bad news for India and China is that their hot economies are causing upward pressure on prices and wages. Wage inflation in 2010 was about 10% driving up price inflation across the broader economy. However, the actual wages paid to employees are still very low so even with a 10% increase the opportunity for labor arbitrage still exists.

Note: The contract contains verbiage covering rate increases. The increase is applied to the hourly rate being charged to the client, not the actual wage paid to the individual. For example, if the offshore resource is billed at $30/hr a 10% increase is $3/hr or $6000 per year. If that offshore resource is paid $10/hr, a 10% increase to the employee is only $1 per hour or $2,000 per year. This seems like a lucrative opportunity however, one must realize that the supplier's overhead and operating costs are also increasing. It is well worth the time to leverage the transparency supplier's offer on their cost structure. It will change how you approach contract negotiations.

Cost structures, exchange rate and monetary stability:

This is a very difficult topic to get a handle on especially given our current global economic climate. Exchange rates tend to fluctuate therefore it is prudent to consider contract language that addresses this issue with some sort of "exchange rate stabilizer clause." Another useful tip is to have all costs within the contract be expressed in USD and let the currency fluctuations fall to the outsourcing partner to manage. To be sure, their rate structure will definitely reflect risks they see with exchange rates but the price to you will remain predictable. Some countries have additional special taxes that may be "added" to what you thought the labor rate was. Be sure to utilize the finance and legal resources on your team to develop price protections needed for multiyear contracts.

U.S. Immigration and work permits:

The most common form of work permit for foreign nationals working in the U.S. is the H-1B VISA. It allows the individual to work in the U.S. for up to six years. The U.S. firm that will employ the foreign worker manages the H-1B VISA process. In most cases the top outsourcing firms have established offices in the U.S. and will therefore manage the VISA process. However, be sure to allow time in your plans for the supplier to secure and process VISA's for resources that will work onshore. Although your partner will most likely manage this they may require a resource commitment from you prior to starting the process. It is also important to note that there are quotas on the number of resources that are allowed to attain work permits. I cannot emphasis enough the need to plan resource needs at least four to six weeks into the future.

Every country is different with respect managing work permits. For example in one case, the Mexican Government required that no more than 10% of the employees working for an Indian firm operating in Mexico be Indian; the rest had to be Mexican nationals. Large multinational firms specialize in creating outsourcing centers across the globe. They invest the time needed to understand the rules of engagement in each geography they plan to expand into.

Time zone issues:

As you define the role your outsourcing partner will play, time zones become important. When it is noon in Chicago, it is 11:30 in the evening in India. The good thing about this offset is that the Indian day shift is the U.S. night shift. This is great for those U.S. employees that had been required to be "on call" to respond to maintenance issues in the middle of the night. Now their Indian counterparts can manage the incident during their day shift.

Unfortunately, many incidents that occur during the U.S. daytime may require the Indian-based resources to work on the third shift in India. Offshore resources covering third shift support may be able to address a limited number of incidents in an "on call" mode. However, for applications used heavily during the day in the U.S. a permanent third shift in India may be needed to assure a rapid response to incidents when they break. For example, when the warehouse management system stops taking orders the technical

Support Team needs to address this as a "Severity 1" incident meaning it must be resolved in a few hours. Just as in the U.S., Indian resources do not want to work third shift. We will expand on this issue in a future chapter.

Chapter 2:

The Case for Outsourcing

Outsourcing IT Services is not a trivial exercise. Essentially it is a decision to trust another firm with your IT capabilities, business knowledge, security and Intellectual Property (IP). Although it can be undone, reverting back to where you were pre-outsourcing (sometimes called "insourcing") is a very laborious and costly process. So, before deciding to outsource, make sure your motives, strategies and plans are well developed. The first question that needs to be addressed when making the case for outsourcing is: "What problem are we trying to solve?" The answer to this question is varied but financial objectives – specifically expense reduction – often dominate. In addition to improving the bottom line, outsourcing can also improve resource optimization, service quality, and provide a more stable foundation for the growth of the IT organization.

We will focus on outsourcing Application Support and Application Development in this book. Development, of course, is the creation and implementation of new software tools or major upgrades to existing software. Support is the group that keeps the lights on. Their role is to make sure every application runs as designed, on time, everyday – and when they don't, they fix it.

Financial:

In many organizations, outsourcing application support is a key element in achieving financial and/or budgetary objectives. IT organizations tend to have large budgets, and therefore stand out as plump targets when corporate-wide cost cutting becomes necessary. Decision makers outside of the IT organization often have very little knowledge of how the IT budget is structured and what drives it. Sadly, many within the IT organization don't have a clue either. For illustrative purposes, let us assume we are working with an IT budget of $100 Million dollars. A budget this size certainly must have 5%-10% of cost reduction opportunities in it – right? Well, not necessarily. Digging down a few layers we can and look more closely at the components of this budget. The budget contains non-discretionary spending categories that cannot be cut as well as others that are very painful to cut. Annual depreciation charges are non-discretionary. They result from capital expenditures for IT development and hardware purchases and amortized for up to seven years. This means the operating budget for each year in the future carries a portion of current year capital spending called depreciation expense. For example, if annual capital spending is $35 MM (35% of the total budget in our example) there will be an ongoing depreciation expense of $5 MM added to each of the next seven years following completion of IT projects.

See Table 2-1.Assuming capital spending is constant from year to year, the sum total of each year's annual depreciation expense will eventually grow to $35MM as each year of amortization gets added to the previous 6.

Table 2-1 Software Amortization

	Year 1	Year 2	Year 3	Year 4	Year 5	Year 6	Year 7	Year 8
Capital Spend >	$35MM	$35MM	$35MM	$35MM	$35MM	$35MM	$35MM	$35MM
Depreciation								
Year 1	$5MM	$5MM	$5MM	$5MM	$5MM	$5MM	$5MM	$0
Year 2		$5MM	$5MM	$5MM	$5MM	$5MM	$5MM	$5MM
Year 3			$5MM	$5MM	$5MM	$5MM	$5MM	$5MM
Year 4				$5MM	$5MM	$5MM	$5MM	$5MM
Year 5					$5MM	$5MM	$5MM	$5MM
Year 6						$5MM	$5MM	$5MM
Year 7							$5MM	$5MM
Year 8								$5MM
Total	$5MM	$10MM	$15MM	$20MM	$25MM	$30MM	$35MM	$35MM

Accounting rules dictate that once amortization begins, there is nothing the CIO can do to stop it. An important realization is that due to these rules, every dollar of capital spending cut in a given year only reduces the current year budget by about 14 cents. A more devastating result can occur when active projects are cancelled (due to spending cuts or a strategy change) all funding spent on the project to that point converts to expense in the year in which the project is stopped. This will have a significantly adverse effect on one's cost reduction plans. A better approach to capital spending for IT may be to charge depreciation expense to the budget of the client that is benefiting from the project. It is easy to endorse frivolous projects when one is not responsible for paying for the ongoing expense generated.

In our example if all projects are stopped in a given year, the savings (expense reduction) on the $35MM of capital is only about $5MM. Unfortunately, the savings will be offset by the immediate expense caused by canceling projects that are "in process" which could be as high as $35MM. Additionally, there would be a cost to layoff seasoned employees and assuming they will be able to find work elsewhere, who knows if they will ever return. Clients often rely on IT projects to facilitate cost reduction programs for their functions. When projects are cut, client productivity efforts will likely suffer as well. Instead of achieving cost reduction objectives, things may actually get much worse from a corporate perspective. Cutting capital spending must be done with involvement and counsel from the IT Finance team and client groups.

Software maintenance fees amount to 15% of the total annual IT spend on average or $15MM in our example. These are fees paid to software vendors for support and future upgrades to purchased software. These are contracted fees, difficult to reduce and should be paid to avoid the risk of using unsupported software. For example, when you buy a software package for $1MM dollars, there is an annual maintenance fee associated with that purchase. Typically that fee is 18% of the purchase price for each year you own and use that software. The software purchase for $1MM results in an ongoing cost of $180,000 each year for maintenance. This allows the IT team to call the vendor for support issues AND receive free upgrades to the software. The software upgrade is free but not the labor to implement it.

When purchasing software, it may be smarter to pay more up front per license in exchange for lower ongoing annual maintenance fees. Example: When I negotiated the purchase of some ERP software the "final" price from the vendor was $2,500 per user plus annual maintenance of 18% of the purchase price. With 500 users and a 7-year amortization schedule, the annual cost for depreciation and maintenance would have been $400,000. I offered to pay them $3,000 per license in exchange for a 4% maintenance fee increasing 1% per year to a max of 9%. They accepted my offer and in the first year our total cost for depreciation and maintenance was $275,000 or $125,000 lower! Over the next 7 years, we saved about $600,000 or 22% vs what we would have paid under the original offer. The vendor was motivated to close the deal at the higher license fee because they were able to post higher sales revenue. Of course getting this done just before the end of their fiscal year provided added leverage as well. The salesperson was motivated too since it meant a larger commission. Remember, the first rule of successful negotiation is to know what motivates the other party.

Application Support on average comprises 35% of the annual IT operating budget or $35MM in our example. The primary cost driver is salary and, because of the nature of work, it is pure expense. Therefore reducing a dollar of support expense yields a dollar saved to the budget's bottom line. Cutting support is a popular notion when IT cost reductions are needed but must be approached carefully.

Support Teams work closely with the users to assure that their applications run on time, as designed, every day...and when they don't, support is there 24x7 to fix it. The user community is not patient when applications stop working. They want their tools to work as reliably as the phone and lights in the office. When support spending is cut, make sure the users are informed of any potential service reductions ahead of time so they know where they may be impacted.

Outsourcing support, when implemented correctly, will reduce costs and improve quality. This is driven by labor arbitrage in remote offshore locations as well as process improvements. Details on how these will be accomplished are explained throughout the remaining pages of this book.

➤ Other expenses such as voice and data communications as well as Technical Services comprise about 10% of the IT budget or $10MM per year. These expenses are difficult to cut as well.

To recap our $100MM budget example, $95 Million is non-discretionary or at the very least, painful to cut. The remaining 5% is made up of management salaries, consulting, travel, and other miscellaneous expenses. When the decision is made to reduce that $100MM budget by 5% or $5MM dollars annually, there are few levers the CIO has at his/her disposal. Where can the cuts occur? Lets be optimistic and assume that $1.5MM of savings is found by reducing capital spending (remember, only 14% of each dollar cut from a capital project is an expense reduction); restructuring to eliminate a few management staff; stopping payment on a couple of software maintenance agreements and cutting travel. Unfortunately, when the dust settles, there still exists a $3.5MM gap that needs to be plugged. Due to the nature of IT budgeting, the next most logical place to look for savings is the money spent on application support.

Cutting the support budget by 10% will bridge the $3.5MM gap. Although it is a lucrative option, at some point cutting support staff becomes problematic. Cutting staff means fewer people are available to address incidents so it will take longer to resolve them and they will start backing up in the queue. This results in lost user productivity and potentially lost sales revenue that may dwarf any savings achieved in the first place. What is needed is a method to reduce the cost per hour of support rather than the total number of hours. Outsourcing support to an offshore partner is an effective way to do just that. It is not a quick fix. An outsourcing initiative will take time to plan and implement. It will also require incremental spending to pay for transition resources as well as severance paid to terminate employees. This is worth repeating. *Outsourcing will require incremental spending to pay for transition resources as well as severance paid to terminate employees.* We will cover the cost benefit analysis of this decision in detail in a later chapter. There are many other benefits beyond cost reductions that can be attributed to outsourcing such as:

Resource Replacement:

Prior to outsourcing, it was a challenge to find and retain critical skill sets in-house needed to manage the technology and applications in the IT portfolio. Hiring from the outside to fill an open position was often met with "No, we are in a hiring freeze" even though I had budget money available. The internal churn caused by hiring a person from another IT function (rather than the outside) was exasperating. Robbing Peter to pay Paul is not only suboptimal, Paul always wins!! Outsourcing changes the paradigm because the supplier is responsible for finding and training replacement resources. My experiences with large partners have shown that a key resource can be replaced quickly and with a minimal drop in capability. To accomplish this, they develop a shadow team of resources (at their expense) creating a backstop for each critical position.

Knowledge Transfer:

One of the more interesting things about working with an outsourcing partner is their strategy of rotating resources every 18-24 months. This is designed to minimize turnover by providing new assignments that help their staff climb the career ladder. When interviewing potential partner firms, ask them to explain their process for facilitating knowledge transfer and experience building within their resources. A common approach is to use junior resources in India to learn the applications and actually participate in incident resolution. In this way, when a rotation into an open position does occur, the junior resource is familiar with the application and better prepared to step into the role making the transition quick and relatively painless. Make sure you agree during the contract negotiations that any transition costs to replace existing resources are borne by the partner. It is common for client staff to participate in "reverse testing" to assure that the new resource is ready and able to step in. If Service Level Agreements (SLA) are being met, the client should not really care *how* the partner manages their talent.

Process Dependence:

The planned rotation of resources every 18-24 months is daunting and not how the traditional IT organizations in the U.S. think. We tend to adopt the "position for life" concept. The organizations I have worked with have people that have done the same job for 5-10 years or even longer! They are very "people dependent" and face a major dilemma (OK, sheer panic) when a key resource quits, retires, or leaves for another job. I have seen managers beg and plead with the individual to stay and/or try to convince potential hiring managers the world will come to an end without their key individual. The real question that begs to be asked is this: "If that key person is so critical to the operation, why has there been no cross-training or other preparations to minimize the impact of their leaving?" Few of us invest the time and money needed to develop a depth chart. Instead of relying on key individuals, top tier outsourcing firms have become "process dependent" instead of "people dependent" through the use of sophisticated knowledge transfer tools. Their philosophy, in other words, is that anyone, with the proper training and knowledge should be able to step in and perform any job with minimal issues as long as they follow the documented processes. Over time, the rotation policy of the partner firm creates a fairly large group of resources with a high level of knowledge about your application portfolio. As future development work arises, these resources can fill important roles on project teams resulting in true resource scale-ability.

Incident Management:

Capabilities can be improved greatly through the use of professional outsourcing partners. Most top tier partners work within an ITIL framework. ITIL (Information Technology Infrastructure Library) is a set of standards used to define common practices, terminology and procedures to optimally manage and resolve service disruptions. For ITIL to be effective, the organization must have a well functioning ticket management system. Ticket management requires a common database to track all incidents. This in turn requires that all users must call the Help Desk and open a ticket to report service disruptions. The ticket details become the foundation for performance tracking and process improvement. Typical statistics include: time to respond to the incident; time required to resolve the incident; root cause; actions required to restore service; etc. Further, these statistics become the foundation for Service Level Agreements (SLA's) with the supplier. SLA's drive the relationship and it is common for suppliers to be penalized if/when they miss SLA targets. Additionally, tracking the root cause of each incident and associated failure analysis leads to small projects and/or enhancements to the code base to improve service quality and reduce the number of incidents in the future. SLA's will be covered in detail later in the book.

A big hurdle when outsourcing support is convincing the user community to adopt the Call Management Center (CMC) and Help Desk processes. Changing a culture is never easy. It sounds like a simple thing to tell the users to "call the Help Desk." The reality is that users and IT staff have worked closely together for years resulting in strong relationships. This can be a good thing but gets in the way when trying to capture statistics on service failures and incident resolution. The traditional way to report an incident may have been for the user to pick up the phone and call Jimmy in the IT department. Jimmy, always anxious to please, jumps on the support issue and tries to solve the problem regardless of whatever else he was doing (i.e. working on a development project, or sleeping, or whatever). The user is gratified and Jimmy is too because he was able to be helpful. Unfortunately, the pick-up-the-phone-and-call-Jimmy process is less than optimal. This is a great example of being "people" vs "process" dependent. Who will the user call when Jimmy is on vacation or home sick or moves to a different job?

Other benefits associated with implementing a formal Help Desk and Call Management Center include:

➢ The user calls one number for help when systems are not working properly.

➢ The help desk operator opens a "ticket" and routes it to the right IT person on call at that time.

➢ The statistics captured by the Help Desk will lead to improvements in service quality and future reductions in operating cost.

➢ Formal ticket tracking provides data that lead to opportunities for a permanent code fix and future quality improvements.

➢ Ticket data can be used to measure operational aspects associated with the SLA's in place.

The company culture will need to change in order to embrace the use of the Help Desk. The battle is well worth fighting.

Although outsourcing is not a prerequisite for a formal Help Desk, the reverse is not true. Don't bother spending the money and time to outsource support without top management commitment to establish a formal CMC and Help Desk. Ideally, the CMC process is in place a year or two ahead of beginning the outsourcing transition. The Help Desk data will be invaluable in determining staffing levels as well as SLA capabilities needed for the contract. A mature Help Desk process should be considered as a prerequisite to any outsourcing project. It is essential for communication and the source for operational data.

Organizational Responsiveness is a term used to describe the ability of an IT organization to quickly respond to key initiatives. Internal staff can provide the deep business knowledge required to make these initiatives successful. Outsourcing support may free up key internal Subject Matter Experts (SME's) to work on core projects allowing the firm to deploy their best resources to core projects. As we noted earlier, core work is what the company needs to do to grow profits and improve shareholder equity. Commodity work on the other hand is important but is not critical to delivering the business plan. Most activities in the application support function are commodity work and therefore can be successfully outsourced.

One of my prior clients was able to repurpose about 50 employees from their support team to work on new ERP implementation. The support work done by these individuals was outsourced to an Indian offshore partner. The financial gain in doing this was phenomenal. First they captured a solid cost reduction by replacing their internal resources on the support team with lower cost labor. Second, there was even a larger savings that came from using experienced employees on the ERP project thereby eliminating the need to hire costly external consultants.

Making the case to outsource application support brings a number of compelling benefits that will enable the IT organization to move forward in a positive direction. I can't emphasize enough the need to properly plan and execute any outsourcing initiative. Starting the journey with application support is a common approach but is more risky than starting with development since support contracts generally span multiple years. One can "get a feel" for a supplier's capability and personality by using them on small, short-term development projects first.

Either approach can work with a strong team in place to evaluate and lead the effort. Plan well and prosper...

Chapter 3:

Defining the Project Vision & Scope

In this chapter we will begin to create the outsourcing "vision and scope" by looking at the four primary functions that are typically outsourced; the various types of outsourcing relationships; and finally the applications and work elements that are best suited for inclusion in the outsourcing scope.

Business Process Outsourcing:

BPO should be managed by the client organization in need of outsourcing one or more of their processes. Although not an IT responsibility per se, the IT organization will be part of the team to help with the transition. Common BPO activities include Accounts Payable, HR, Employee Benefits, Order Capture, and many more. Using Accounts Payable as an example, we find that although the payment processes can be automated, many lower level employees are used to review accuracy and resolve customer issues. Today's web-based technologies allow this type of "commodity" work to be easily moved to low cost geographies anywhere in the world to capture labor arbitrage savings.

Application Development:

The Application Development function is responsible for creating new software solutions to resolve business problems. The methodology and focus is to work closely with the business clients to gather requirements, construct the application, test the code and validate that the client got what they wanted. Coding and testing take should always take place in an isolated environment with no direct connection to active code. Once the new code is developed and tested it will be moved to production. If a development project is delayed, there is no immediate adverse impact on the business. The adverse effects from project delays are not nearly as devastating to the business as when active applications stop working. Funding is primarily capital amortized over a five to seven year period.

IT Infrastructure:

This function manages hardware, operating systems, data centers, telephony, communications, security and networking. These activities form the backbone upon which the enterprise will process and access data. This area typically consumes the largest portion of the IT budget. It also draws the largest crowd when incidents occur. A server crash or loss of telephone connectivity creates an immediate impact on business users.

The funding in this area is a mix of capital (to purchase hardware) and expense to manage it. Strategic planning and execution balances the long-term viability of the organization against the daily operation of the enterprise. Without a long-term vision, "network" viability will deteriorate as new versions of technology demand more processing capacity. This team works closely with Development and Support to plan and deliver the environment changes required.

Application Support:

The Application Support function makes sure that every application works as designed, on time, every day – and when they don't, they fix it. The work done in the Application Support function lends itself very well to outsourcing and is therefore a common place for firms to start their outsourcing journey. Funding for support is primarily expense and, as mentioned earlier, is a prime target when cost reductions are required. Application Support is vital for the day-to-day "running" the enterprise. Throughout the course of the book we will refer to it as the common frame for much of our discussion in this book.

Supplier Concentration:

Many of the top outsourcing partners can provide most of the services in these areas. Often, the desire of the client is to divide the work across multiple partners rather than using one or two. This practice may minimize the presumed risk of "putting all of your eggs in one basket" but could leave some benefits on the table. Arguments in favor of bundling services under one supplier include: it is easier to manage a single supplier vs many; potential for better pricing; synergy opportunities between functions are easier to achieve; less finger pointing when incidents occur. Arguments in favor of spreading the risk to multiple suppliers include: fosters competitive bidding for new work; built in "back-stop" in case one of the suppliers needs to be replaced and it minimizes the likelihood of a single supplier holding the IT organization hostage. The Executive Team should evaluate the alternatives and formulate "guiding principals" regarding which approach is appropriate for the IT outsourcing charter.

I have found that a single supplier within a function offers a much better solution than spreading the work to many. For example, incident resolution in the Application Support function can span many application areas. A ticket opened by a supply chain user may touch Sales, Finance, and/or Procurement applications. With a different supplier providing support in each of these, the flow of communication deteriorates and fingers start to point. None of the suppliers want their SLA to take a hit. Another reason to bundle under a single supplier is the ability for that supplier to build a solid knowledge base about the applications they are supporting and how the client's organization functions. Admittedly, this depth of knowledge positions them to better assess what it takes to be successful at winning future bids. As more and more work goes to the primary supplier, other suppliers wise up and decline the option to bid on future work. They realize they are being used as a negotiating "wedge" with no hope of ever being awarded the work. To avoid this, I suggest that the entire functional area (domain) be put forth in the initial Request for Proposal (RFP). The RFP can then be sent to multiple suppliers for competitive bidding rather than taking a piecemeal approach. In other words, there would be one RFP for the entire Application Support function, one for Development, one for Infrastructure and another for BPO.

There are various forms IT outsourcing relationships can take. The focus of *The Inside of Outsourcing* is the "Managed Services" model, but it is important to take a moment to explain the three basic ways IT services can be delivered.

Consulting:

Consulting engagements involve the short-term use of third party expertise to help solve an immediate business problem. Resources with significant skill depth in a specific domain will be called in for a short duration, provide a solution and then leave. IT consulting has grown in recent years driven largely by the rapid growth of new technologies. The rapid growth and changes to existing technology makes it difficult (if not impossible) for an IT organization to be "expert" in everything. Consultants provide a vital role filling the knowledge gap by bringing expertise to help the organization learn and adapt to new technologies. Good consultants are worth their weight in gold. They can produce the desired output in a much shorter period of time than can an inexperienced staff.

The hourly rates for consultants can be very high. One of the reasons for this is the instant expertise they provide to fill gaps in the organization's knowledge base. The old adage "knowledge is power" is quite true and in the case of consulting, knowledge is money. Consultants fill a broad range of needs from helping with organizational design, to implementing an outsourcing project, to providing specific application knowledge (often the case with major ERP implementations). Another reason for the high hourly rate is that engagements tend to be unpredictable and are usually short duration. To assure availability when a client calls, the supplier will maintain a "bench" of available experts that may not be busy 100% of the time but, to retain them the supplier needs to pay them 100% of the time. Higher rates are required in order to make payroll during low points in their revenue cycle. Supply and demand also impacts the price.

Here are a couple of ideas to help minimize the overall cost of a consulting engagement. First, prior to engaging a consultant, be sure to document and frame the problem you want help with. This reduces the time needed to bring the consultant up-to-speed. Second, make sure internal staff is available as needed to supply input and background to the consultant. This will help minimize the number of billable hours charged. Finally, when negotiating a consulting agreement, a better rate may be available for a longer commitment or a renewable engagement.

Staff Augmentation:

Staff augmentation is quite literally what it sounds like. The IT organization is augmenting its in-house staff with third party resources. These resources tend to work as "replacements" for the traditional IT employee. An insufficient number of internal resources can result from layoff, attrition or simply because hiring has not kept pace with the appetite for IT projects. Corporate logic says that renting IT resources from the local purveyor of talent makes more sense than hiring. Initially, the work they do is "context" in nature, typically coding and testing. They are closely monitored and directly managed by employees within the IT organization. Over time they become Subject Matter Experts (SME) for the applications they work with. If left unchecked, they will eventually become one of those "key resources we can't live without" discussed in Chapter 2. If allowed to get to this point, not only does the IT organization suffer by becoming more "people dependent" but can be held hostage by firms that supply those resources. The rate per hour for staff augmentation resources can easily be 1.5 to 2 times higher than the cost of the internal staff they are replacing. Many of the temp agencies providing staff augmentation resources do not provide onsite management nor do they invest in developing their resources. The agency simply manages an inventory of skilled staff that is deployed in response to client requests for specific skill types.

Staff augmentation can be a very effective way to provide a flexible pool of resources as project demand fluctuates. However, problems can arise when they spend long periods of time (years in some cases!) within the IT organization. They become an integral part of the fabric. A long engagement with staff augmentation resources quickly becomes a slippery slope that the CIO needs to be concerned about. Yes, the CIO needs to be on top of this! The problem starts when project demand exceeds the available number of internal staff needed to deliver them. One project leads to another, and another, etc. The IT management team (always wanting to please the clients) keeps going to the staff augmentation "well" over and over to deliver projects. They are more focused on delivering projects than with the general health of the IT department. Eventually a strong dependency develops with the staff augmentation resources. Fear grips everyone's heart and panic sets in (especially with mid-management) when that "key" resource leaves because they take with them a depth of knowledge not easily replaced. Sadly, those left behind may have lost, or in some cases never had, the requisite knowledge required to effectively manage the applications.

Unfortunately, heavy reliance on staff augmentation for long periods of time not only decays the health and strength of the IT organization but can also lead to legal issues as in the case with Viscaino vs. Microsoft. In this landmark case long-term temporary workers claimed (and won) they were due certain benefits from Microsoft. Here are a few suggestions to insure that staff augmentation does not become the extremely addictive "drug of choice" for your organization.

➢ Enact mandatory 60-90 day breaks in service for staff augmentation resources after 6-12 months of continuous service. This means that internal staff will need to keep their application knowledge intact, which is a good thing. These breaks in service will add to the cost of the staff augmentation model due to the productivity lost when replacement resource are trained. Also, it is rare that the same SME will be available after their 60-90 day break, as top-notch resources will be redeployed to some other client. IT managers will spend endless hours trying to figure out how to work around this hurdle. Don't let them!!

➢ Request that the HR team researches the terms of service with each contractor and conducts regular reviews with the IT management team. Force them to identify the SME's (internal and external) and work to reduce the list of those that are staff augmentation resources. It may be better to hire third party SME as an employee (assuming the agreement with the vendor allows for it) than to continue using them in a staff augmentation role. This is a common practice and actually reduces risk as well as costs.

> Create Knowledge Management protocols and conduct regular reviews with staff to assure these protocols are being followed. Creating a Knowledge database will help assure that internal staff retains their application knowledge and will accelerate the learning curve for new resources.

Managed Services Model (MSM):

Managed Services is defined as moving a defined work segment to a third party partner that provides management oversight and direction to their resources. The client and supplier need to define Service Level Agreements (SLA's) and Key Performance Indicators (KPI's) together. These will be used to manage the effectiveness of the third party. For example, if all support activities are performed by the supplier, the client will want to know how well they are doing against key metrics such as "Time to Respond" or "Time to Resolve" each incident. The client IT staff no longer performs the day-to-day work and transforms from "doers" to "reviewers". As reviewers they rely on SLA's to govern the supplier making sure they deliver desired outcomes. There are five key enablers outlined in Table 3-1 that will improve the chances for a successful move to the Managed Services Model.

Table 3-1 Managed Services Model (MSM) Enablers

Strategic Concerns	Tactical Focus
Demand Management	- Centralize the IT project portfolio - Establish planning rigor - Balance funding and resources
People	- Change the way SoW's are written - Change from "doers" to "reviewers" - Think and act like "owners"
Knowledge Management	- Capturing existing and new knowledge - Document solutions for known issues - Record foundation knowledge - Storage and retrieval of artifacts
MSM Supplier Partners	- Identify a strong supplier partner - Consider long-term agreements - Manage with SLA's - Evaluate Supplier Relationship Management (SRM)
Infrastructure Partners	- Alignment and process discipline - Network reliability - Service planning reliability

Demand Management:

Demand Management is a formal centralized process that collects, approves and prioritizes all IT projects across the enterprise. Too often, IT organizations are merely a conglomerate of fragmented systems teams (Finance, Supply Chain, HR, etc.) with each of them maintaining a project agenda for the IT resources they "own." Although popular with the business partners, this can be an inefficient way to spend the IT dollar. Project funds for one function (HR perhaps) may not deliver the same benefit as spending the money on a project with a higher overall return for the stockholder. The project portfolio becomes the complete list of all IT projects prioritized by strategic importance and financial benefit to the enterprise. Spending decisions are made from a portfolio perspective. The portfolio plan will not only drive IT spending but also help Development Teams plan and align key resources across all IT functions required to implement the project (infrastructure, support, master data, security, Quality Assurance, etc.).

Another benefit of Demand Management is improved planning. In the old model, management and oversight roles were replicated in each function and any common practices and standards happened by chance rather than by design. The ability to manage projects spanning multiple functions became increasingly difficult as well. With Demand Planning in place, the team is in a position to recommend solutions and tools to facilitate resource planning through the use of common development protocols. The team will be able to monitor and assure that time tracking is complete and project milestones are reviewed on a routine basis. They may also have the authority to withhold funding until requirements for each phase of a project are complete and approved.

Funding is not the only constraint to project development. There is also a need for capable resources. I have seen it many times; the end of the year is approaching and unused capital becomes available. The expectation of upper management is that IT resources sitting on the bench just waiting to hop up and efficiently convert the newfound capital into a new application. I guess this happens on occasion but for the most part, nothing could be further from the truth. The need to balance funding and resources is important – too much of one vs the other results in problems. The Demand Management team is in a position to keep these two elements in balance.

People in the IT organization need to transform how they work. They need to think in terms of "outcomes" rather than tasks. The traditional IT employee was groomed to be task oriented. Lines of code cut without error was the traditional way of measuring one's effectiveness. However, delivering business value has become the new way to work. Today's IT professional needs to focus on how business outcomes are delivered through applications.

For example, in the coffee business, the flavor of the brew results from a blend of Arabica and Robusta beans. Complicating matters is whether the beans come from West Africa, Brazil, Hawaii or Indonesia. The depth of flavor is a result of the level of roasting the beans undergo before grinding. The price of green coffee beans is market driven and changes constantly. The business partner may need to know the right mix of beans needed to do two simple things: create a finished product that meets consumer desires for flavor while minimizing the cost of materials. The challenge for the IT staff is to create an application with a Linear Programming (LP) engine capable of incorporating the thousands of variables and hundreds of constraints involved in creating the optimal coffee formula. The formula is the "outcome" the business needs to solve their problem and this outcome needs to be defined in the SoW. The old way of writing a SoW was to simply order up six Java coders (no pun intended) and two testers.

In the MSM model, internal staff is no longer directly involved with coding and testing and will change from "doers" to "reviewers". They need to monitor the supplier to assure that project milestones and outcomes are met. Also, we (their managers) need to give them the autonomy to step up and manage MSM as their own business.

Knowledge Management:

Knowledge Management (KM) assures current and future knowledge about the applications is captured, stored and retrievable. This should be started well before letting people go. The supplier will help with this during the Knowledge Transfer process and use what they learn to fill gaps. Part of the supplier's duties ongoing should be to record each incident and the resolution so in the future one can simply scan a database to look-up an incident and how it was resolved.

Knowledge Management involves archiving application and technical documents for future use. Some of these may already exist and some will need to be created. A Knowledge inventory should be built for each application so the organization knows where deficiencies exist. Context diagrams, Application Information Documents (AID's) and others must be stored in a central repository that can be accessed by development or support resources that may be in need of them. These can be used for training future resources assuming they are kept up-to-date. I will say it again; all documentation must be Up-To-Date or it quickly loses value. We will discuss governance roles in a later chapter but one of those roles is to assure the supplier keeps the Knowledge Management database current.

Outsourcing means that 80-90% of the existing staff will either be exited or redeployed. Gone with them is knowledge about the applications, environment and business processes collected in their heads over the years. Don't let the zeal of capturing cost reductions upstage the need to create a comprehensive Knowledge Management program. Sadly, the severity of the loss is rarely anticipated and not experienced until it is too late.

MSM Supplier Partners can facilitate the transition and help assure success. Not all suppliers are capable of doing this however, so be pragmatic when you select a partner. World Class suppliers have developed training programs designed to help the client migrate to an effective Managed Services Model.

Suppliers need to invest in building their knowledge about the client's IT organization, the application suite, and the technical environment. This doesn't happen over night. In order to recoup their investment, a supplier will be looking for a multi-year agreement especially for an application support engagement. Over the years, IT organizations realized that the best way to learn was to start on the support desk. The same is true for supplier partners. Managing application support affords the supplier a close-up view of the applications, how they relate to each other as well as how they fit in the broader technical environment. The natural migration is from the support desk to work in Application Development. The environmental knowledge gained from support gives the supplier an advantage over other suppliers when bidding on new work.

Supplier Relationship Management (SRM) for IT services is another enabler of MSM. The SRM team manages the procurement process with all technology vendors. Chapter 10 is dedicated to SRM.

Infrastructure Partners are an integral part of any IT organization. Keeping Development, Support and Infrastructure Teams aligned against common objectives helps enable MSM. The Infrastructure team is also responsible for installing and maintaining the connection between the client's office and the U.S. hub used by the supplier. With MSM, 70-80% of the project resources will be located offshore and when the connection is lost, those resources are in the dark.

The Infrastructure Team can impact the delivery pace of projects through their service planning function. Development projects need environments to work in, test in and operate in and are typically under the purview of the Infrastructure Team. Reliable service planning should be able to accurately commit to a date when new elements will be set and operational. This helps the Development Team plan and minimize the cost of project delivery. There are few things as frustrating as hiring development staff and having them burn dollars while sitting idle waiting for the environment to be set up – it happens far more often than it should. Successful development, regardless of which flag the resource carries, requires reliable service planning.

Applications and Work Elements:

Not every activity lends itself to outsourcing. Successful outsourcing starts with outsourcing the right type of work. Table 3-2 describes various types of work along with the likelihood of attaining the benefits of outsourcing. Table 3-2 shows that the highest success comes from outsourcing the type of work that can be completely within the control of the supplier. This is generally what we have defined as "context" work.

Context work is important but the person doing it should not need to seek client approval for every action they take. The work follows specific procedures and lends itself to script-driven processes that minimize ambiguity and rely on explicit data and solid documentation. Explicit data is that which is precise, clearly defined, categorized and searchable. An example of explicit data might be the instructions for resetting a password. Application Support and incident management fit well in the far left column on Table 3-2.

Call center work is also well suited to outsourcing which is why so many firms have moved their call centers to India. Listen to the person you are talking to the next time you call any sort of help desk (insurance claims, cable providers, credit card inquiries, etc.). You are likely to detect that they are following a script. Get them off their script and you are doomed…let them follow the script to increase the chances of getting your issue resolved. Not to sound condescending, call center staff are not typically high-end technical or business process wizards. They need to have a good demeanor, a good command of the user's language and the ability to read and follow scripted responses. Processes that are specific and well scripted can also be automated and reduce costs.

The far right side of Table 3-2 is rarely (if ever) a good place to install a Managed Services solution. The work done in this area is generally referred to as "core" work. It typically includes large projects that often require the business to change how they do what they do. This is, however, a good place to consider a consulting engagement to bring in expertise that will make recommendations for management to review and approve. One of my prior employers went through a period of multiple acquisitions resulting in a mishmash of order and fulfillment systems. The VP of Distribution questioned the long-term viability of the resulting network of Distribution Centers (DC's) and brought in a consulting firm to study the network and make recommendations to drive efficiency and improve service to our customers. What evolved was a multiyear strategy essentially resetting the entire network of physical DC's as well as implementing new systems for order processing and warehouse management.

Table 3-2 Managed Services Success Drivers

What type of work is best suited to a Managed Service relationship?

Highly Successful	Sometimes Successful	Rarely Successful
Entirely Within Supplier Control	**Requires Business Approval**	**Requires Business Behavioral Change**
Work that can be done by the supplier with minimal client involvement such as CMC, Incident Management, Minor Coding and Testing	Work that can be initiated by the supplier but needs client approval such as code changes and process improvements.	Supplier recommended needing business practice change such as a complex application upgrades or game-changing projects
Requirements: • Script-Driven Responses • Coding Documentation • Explicit Data	**Requirements:** • Application Knowledge • Technical Knowledge • Procedural Knowledge	**Requirements:** • Ability to change how clients do their work • Define CORE business objectives

We have reviewed the various forms of outsourcing engagements and the type of work that lends itself to success in a Managed Services relationship. The next step for the Working Team (WT) is to identify, in detail, all of the applications and work activities that will be "in scope" for the outsourcing project. It is critical that the specific work elements be identified for two reasons. First, labor cost and other expenses associated with performing these activities establish the "current cost" (what is being spent pre-outsourcing) needed for financial analysis. The second reason for determining the detailed list of work elements is that it helps the outsourcing partner understand exactly what it is they will be doing. The partner will validate these activities during their "due diligence" review with the Working Team to assure alignment.

What's in and what's not:

It would be a very rare situation in which every application in the IT portfolio is a candidate for outsourcing. Not only does the Working Teams need to decide which applications make sense to outsource, and recommend which applications are better left as "employee managed." A review session with the Executive Team is timely once the list of "what's included" and "what's not" has been generated. Future planning actions (resource levels, cost, SLAs, contract inclusions, etc.) depend on it.

Employee Managed Applications and Security:

There will be a subset of applications that are best left managed by internal staff. Examples could include applications containing sensitive data, applications that require significant interface with the users during normal workday and new applications that are not yet stable enough to outsource.

Examples of applications touching sensitive data or Intellectual Property (IP) may have databases containing documents with active or pending legal actions, product formulation, market expansion plans, credit card information, etc. The natural reaction from upper management is that third party partners should not have access to sensitive data. This is based on the premise that internal employees are more trustworthy than external resources. If this were a valid concern, it would rule out most applications from ever being outsourced. When the topic comes up (and it will) I like to ask "what makes an employee more trustworthy?" After all, the newspapers are filled with stories about disgruntled employees that hack into their company systems and download IP or cause other harm. The right discussion should be focused on whether or not the controls employed by the firm and its outsourcing partner are adequate to minimize the threat of both accidental and deliberate intrusion and theft. These controls should be consistent with or exceed what the client has in place. Examples of sound security protocols involve both personnel and technology:

Personnel:

➤ Everyone needs a unique account and password.

➤ Establish solid checkout and check-in procedures.

➤ HR policies must describe clear and direct consequences for any violations including termination.

➤ Perform background checks on all incoming employees.

➤ Limit access (by user ID) to the data the vendor will need.

➤ Create procedures for contractors and trusted partners.

Technological controls:

➤ Check contractor equipment for compliance and disable access to USB, CDRW, and DVD-RW drives.

> Encryption software for sensitive data and log security events looking for signs someone is trying to break in.

> Watch for large transactions off the network that may indicate the existence of off-hour data transfers.

> Periodic checks of Payment Card Industry (PCI) compliance and Sarbanes-Oxley (SOX) audits.

A thorough review of security procedures and policies with any potential partner is critical. Make sure your internal legal and risk management auditors are involved and insist that they schedule periodic reviews of security procedures and policies on an ongoing basis. I have found that the larger outsourcing firms have well defined processes supporting internal security especially where client data may be at risk. Any impropriety caused by the outsourcing firm would create huge financial liabilities for them due to lawsuits from the harmed client. Also, and probably more important, is the loss of future revenue due to the extreme negative PR a breach in security would cause for the supplier. Despite their deep pockets (or maybe because them) the supplier simply cannot afford to take any risk that could harm their reputation and future earnings. How much does a disgruntled employee have to lose?

Time Zone concerns:

Time zone differences between the U.S. and India create special advantages and opportunities for application support. Many IT activities such as data purges, job updates, scheduled recycles, etc. occur during third shift in the U.S. These are great activities to perform offshore rather than use scheduled third shift or on-call resources in the U.S. When it is midnight in Chicago it is 11:30 am in India. However, there is reason to be concerned when significant contact between users and the support desk occur during normal work hours. When it is noon in Chicago its 11:30 pm in India thereby requiring a third shift crew in India to manage these calls. Employees in India do not like working on third shift any more than they do in the U.S. The Working Team needs to discuss how the outsourcing partner intends to staff to cover the needs for daytime support.

What I have seen happen is the normal daytime resources in India will be rotated and stretched to cover the third shift needs. This may be okay for the short term, but eventually fatigue sets in and turnover becomes an issue. Here are three suggestions on how to manage this:

> Offshore: As more and more IT work is transitioned to the

offshore partner, things tend to work better. One thing that benefits from more mass is staffing a third shift in India. There will be less of a strain on a 200-person offshore team than one that has only 20 resources. With a larger team, the outsourcer has more people across which to rotate the off-shift work.

➢ Nearshore: The issue is not that the support resources are in a remote office but rather the time zone differential. One answer is to consider positioning part of the Support Team in a country that is closer to the U.S. time zone. Many global outsourcing firms have centers in Mexico or other Latin American countries where the time zones line up better with the U.S. The hourly rate may be higher than that paid for resources based in India but will be a bargain compared to paying onshore U.S. rates.

➢ Some suppliers have developed a "follow the sun" approach to incident management. The incident is passed from center to center following the sun as it traverses the globe. For example, there may be a team in India, a team in the U.S. and a team located in a time zone in between – Eastern Europe is a common location for this. This approach can initially create minor coordination issues for your partner, but they should be able to manage through this. Be sure to test this on a small scale before fully adopting it.

○ Note: it is important that the contract has provisions to restructure the rates and overall costs when different geographies are used. This is critical when a "fixed price" payment strategy is in place. Without this provision, the fixed price will not change even though the supplier moves to lower cost geographies.

➢ Onshore: Position the support resources needed for the high daytime call volume in the U.S. This solution is okay from the perspective of providing the best service to the users. However, it comes with a tremendous cost. Average on-shore rates for third party resources will run two to three times higher than offshore rates in India depending on skill sets and geography. The average fully loaded rate per hour for an internal employee is about 50% higher than the rate of an offshore resource.

➢ As a last resort, the application can simply be taken out of scope. If there is no unique hybrid approach that the supplier can offer to fit the situation, it may be better for this application to be taken out of scope and supported by internal staff.

Other applications that may be better supported by internal staff include those that are still "work in process." Often, new applications are not stable enough to add to the scope of the support contract until many months after implementation. New code is rarely installed without some sort of modifications needed. The cause for this is usually a combination of inadequate testing (especially as it pertains to integration with other applications) and incomplete requirements defined by the user. The incidence rate for new applications is very high at start-up and gradually drops off after a few months as application code fixes and missing functionality are installed. It also takes a while for the user to work through the learning curve for a new application. Whatever the cause, be sure to wait a few months after implementation before formalizing support staffing with your outsourcing partner. Formalizing means adding the application to the list of "in scope" applications defined in the contract. Staffing for application support is driven in large part by the number of help desk tickets. Taking a snapshot of the ticket count before things settle down will result in a larger number of resources than would be required once the new application is stable. Changing the resource count in support contracts is not easy. It is better for both the client and the outsourcing partner to wait a few months and get it right than to staff it wrong and try to correct it later. It is not unusual for third party resources, working in a staff augmentation model, to manage support calls during the post-install period, especially if they were part of the team that built the application. However, this work is usually done on a time and materials basis outside of the fixed bid constructs of the formal support contract.

Once the work elements and all "in-scope" applications have been identified, an estimate of the person-hours needed to complete the work can be calculated. This estimate will become the basis for determining the number of resources needed for the project as well as the cost. For example, when support work is outsourced, the primary work revolves around incident resolution or what is commonly called "Break-Fix." As the scope is defined, one should also make sure that other activities performed by the incumbent Support Team are included as well. Activities such as making minor code enhancements, application testing, release management, Sarbanes-Oxley (SOX) testing, annual application upgrades and other activities need to be specified in the contract rather than assuming they are included. Typically an individual works on multiple activities throughout the day. Many IT organizations utilize time tracking systems with a database that will be invaluable for this analysis. However, one needs to validate the data to make sure what has been reported is actually happening. Too often, the IT employee will simplify reporting and not expend the effort to accurately record the multiple unique activities they may have worked on.

The Working Team will be able to calculate (approximate?) the amount of work hours required to perform the support activities for the applications in scope. They need to consider seasonal spikes in demand for events such as Year-End processing, annual benefits enrollment, Quarter end for warehousing, etc. Lets say this comes to 40,000 hours per year. This would suggest that approximately twenty resources would be needed to perform the in-scope work. The best way to identify all of the work activities is to start by going to the source and meeting with internal staff that perform and understand the application support function. The support staff will know about the "back room" activities that are needed day-in and day-out to keep the applications running as designed. Taking the time to accurately gather all of the work requirements will help to minimize disputes and ill will in the future. Be sure to utilize the experience of the top-tier outsourcing firms as well. Their business is to create an effective outsourcing relationship and can be very helpful with the process. They will review statistics for the applications in question and also provide insight on the multiple technologies involved. Obviously it will be awkward to involve a partner before the contract is signed. Make sure to allow the time (and some leeway in the final staffing numbers) for them to perform a "due diligence" on the work elements involved once an agreement in principal is reached.

Determining the number of Full Time Equivalents (FTE's) that will be outsourced is a very complex exercise. It is generally the case that employees in IT organizations are salaried and not paid for off-hour work. When they are called to work on a support issue in the middle of the night or required to work overtime on a project or come in over the weekend it is assumed to be "part of the job." Employees are rarely paid for the "extra" time they spend. Most outsourcing partners however, will staff to cover any off-hour work that occurs on a routine basis. Off-hour work (whether paid or unpaid) done by employees needs to be factored into the cost analysis for a true apples-to-apples comparison. Here is why this is important:

In the U.S., the average salary for an IT resource in 2011 with two to four years of experience was $68K per year plus 40% for their benefits resulting in a fully loaded cost of $95K per year. If an employee gets paid time-and-one half per hour for five hrs of off-hour work per week, the actual hourly rate (taking into account paid holidays, vacations, sick time, etc) is about $56/hr. If the employee is not paid for that extra 5 hrs of work, the rate drops to about $50/hr. The cost per hour for an internal resource is used to establish the base for our current costs. We will explore this and other cost comparisons in greater detail in a later chapter.

Note: These are typical salaries from 2011.

Chapter 4:

Third Party Labor Models

This chapter will compare and contrast the various types of labor models commonly used to deliver IT services. We touched on the Managed Services Model (MSM) and Staff Augmentation (SA) in the prior chapter. There is a place for both, but which to deploy merits further discussion. Please refer to Table 4-1 as we compare and contrast these two staffing models more closely.

Table 4-1 Comparing Staff Augmentation to Managed Services

Topic	Staff Augmentation	Managed Services
Type of Work	Context work such as coding, testing and support	Context work such as coding, testing and support
Managing	Client manages day-to-day work of the third party resources	Supplier manages day-to-day work of their resources
Training	The client trains the third party resource on an ongoing basis	The supplier is responsible for ongoing training of their resources
Quality	Quality can be high but limited but limited to the scope of the project	High Quality with a broader scope that grows over time
Value	Blended rate will be higher due to a lower ratio of offshore resource ratio of 30-40%	Better blended rate due to higher offshore ratios exceeding 70%
Contract	Designed to protect the supplier cash flow by addressing rates	Defines specific outcomes and metrics to be achieved by supplier
Results	The client is responsible for results since they manage the work	Supplier is committed to the delivery of outcomes
Viability	High risk and high cost associated with using Staff Augmentation	Best for long term partnership between supplier and client
Which to use	Good for short duration projects with defined start and end dates	Best for a permanent change to the IT staffing model
Risks	- Concerns with co-employment - No performance metrics or SLA's - Higher cost over the long term	- Will fail without true partnership - Difficult to press the undo button - Requires long term commitment

Type of work:

In Chapter 3, we explored in detail the type of IT work that has the highest likelihood of being outsourced successfully. Context work (coding, testing, etc.) with well-defined procedures and explicit data is best as it provides the supplier with the ability to make independent decisions and deliver acceptable output. My observation is that the type of work being outsourced is independent from which working relationship exists between the client and supplier. In other words, MSM and SA are both fine for performing context work. To determine which model to use, one needs to compare them side-by-side.

Managing resources:

Staff Augmentation is a process whereby a company brings in external resources that work alongside existing staff to perform specific tasks. The firms that supply these resources act as brokers deploying resources to satisfy requests for certain skills (ie. a Java coder). At best, they maintain a skills inventory for each resource but they rarely do anything in the way of training or managing career path. An example would be a restaurant in need of a temporary grill chef to help out during a busy time of year. The agency would find an experienced grill chef in their resource database and send one to the restaurant. The restaurant staff would manage and oversee the activities of the temporary chef. In the same way, someone in the IT organization is responsible for managing the day-to-day work done by the external resource. It is common to see third party resources working alongside a contingent of client resources with minimal differentiation. They attend meetings, socialize with employees and basically become part of the office fabric. IT management needs to understand the inherent risks when these boundaries blur and evolve into the type of "co-employment" relationship discussed in Chapter 3.

By contrast, in a Managed Services Model, co-employment issues are rare because the third party performs the entire work process with all resources managed by the supplier. Using the restaurant example, the entire kitchen staff would be outsourced to the supplier. Every position from head chef to dishwasher would be a resource working for the supplier. The supplier would be responsible for managing their resources to deliver meals that meet or exceed specific quality and cost standards (outcomes) as defined by the restaurant owner.

Training resources:

With a Staff Augmentation model, resources are expected to have technical skills (i.e. how to code in Java) but the client needs to train them on procedures, specifications and project details unique to the client's organization. To minimize co-employment concerns some organizations strictly enforce stay limits for externals. The problem is that these firms need to retrain Staff Augmentation resources every 6-12 months. Retraining is not only costly but results in lost productivity while new resources climb the learning curve.

In the Managed Services Model, the supplier is responsible for training replacement resources and can be held accountable for meeting predefined SLA's and may be penalized when they miss. Of course the client is encouraged to observe and even participate in the training process to assure the proper knowledge level of the replacement is reached. The larger outsourcing firms I have worked with will actually prepare potential replacements in anticipation of turnover. This is typically done with offshore staff shadowing the primary resources a few hours each week to learn the job. This practice not only grows "replacements" to assure a ready stream of resources to quickly step into future openings, there is no added cost to the client. This is one of the ways Indian firms invest in their employees to broaden their knowledge and prepare them for advancement along chosen career paths. Although there must be one, I am not aware of any U.S. based IT organization that invests in training their employees in this way.

Product quality:

Solution quality is more dependent on the skill and knowledge of the resources than it is on the relationship model being used. Although multiyear Staff Augmentation relationships are common, this model is best used for short-duration projects that have specific start and end dates. The model can produce high-quality results but due to the short duration of the engagement will often be limited to the scope of the project. The longer-term benefit of the Managed Services Model (MSM) is it creates broad exposure to the client's applications and infrastructure leading to superior results.

Value for the money[1]:

The actual blended rate per hour will be lower in a well-executed MSM relationship. Use whatever rates you wish. For analysis purposes, I will be using rates of $30/hr offshore and $75/hr onshore, resulting in a blended rate close to $40/hr. This is driven by the fact it is much easier to achieve a higher percentage of offshore resources with a Managed Services Model. Achieving 75% of the resources working offshore is not uncommon and the ratio of onshore to offshore can actually be "baked" into the contract terms. With the supplier heavily involved with the delivery of Managed Services they take responsibility for achieving high offshore ratios. Note: It may take 12-18 months to reach the agreed targets.

---

[1] *The rates used in this analysis were typical for 2011*

Many Staff Augmentation suppliers use offshore resource too. Since they are usually not bound to reach a predefined ratio, the percentage of their resources working offshore is much lower than with MSM. Again, using the $30/hr and $75/hr assumption, the best I have seen is about 40% offshore resulting in a blended rate of $56/hr. If they use no offshore staff the cost would be $75/hr. This is why the use of Staff Augmentation will cost $32,000 - $70,000 more per resource per year.

Contractual relationship:

The basic difference between the two models can be defined as "involvement" versus "commitment." The old story about the chicken and the pig providing ham and eggs for breakfast is a relevant example. In the delivery of breakfast, the chicken is "involved" whereas the pig is "committed." With Staff Augmentation the supplier is involved to the extent that they match an individual skill to the needs of the client. Staff Augmentation resources follow instruction and oversight provided by the client. The contract for Staff Augmentation is very limited and is more concerned about protecting the cash flow of the supplier than it is about providing outcomes for the client. It will include hourly rates along with financial penalties and/or fees if the client hires one of the supplier's resources. There is little verbiage dedicated to performance, SLA's, outcomes or quality.

The Managed Services contract defines the supplier's commitment to the relationship. In addition to rates, there are: defined outcomes with specific metrics; definition of scope; cost reduction targets; included work activities and more. We will review the elements of a MSM contract in depth when we discuss Supplier Selection later in this book.

Responsibility for results:

When you buy a steak at the grocery store, there is some assurance that the steak meets certain standards of freshness, tenderness and flavor. However, the store is not responsible for how you cook your steak. The supplier in a Staff Augmentation relationship assures their resources have the experience, training and skills required to meet the specifications of the client. However, just like the grocery store, they are not responsible for the work done by those resources. The client manages the resource assigning work and monitoring progress. The client is responsible for overall results.

With Managed Services, the supplier agrees to deliver specific outcomes defined by the client and may face penalties if they don't. The supplier assigns work to their resources, manages results and is responsible for the outcome in the same way a restaurant is responsible for delivering a perfectly cooked steak to your table.

Performance metrics, or Service Level Agreements are used to govern the supplier and assure they are achieving *your* objectives.

Long-term viability:

Staff Augmentation in my opinion is not a good solution for the long-term. Unfortunately, it often becomes the only solution especially when staff levels are cut and hiring is frozen. The business still demands to have their projects delivered! Corporate decisions can be perplexing and, well, just plain stupid. Here is what I mean: Someone at the top makes a Draconian decision to cut headcount across the board by 10%. To deliver this cut process should change (doing things more effectively) and/or there will be a reduction in services. The reality, however, is that processes are rarely redefined and the reduction in services rarely happens. The functions continue to work as before but with fewer people. The workload hasn't changed but there are fewer people doing the same amount of work as before. To make matters worse, there are assumptions made that revenue creation will be the same or better and costs will be reduced. But wait a minute; in order for other departments to achieve their cost reduction and/or revenue growth objectives, they may need new system tools or significant modifications to the existing ones. Unfortunately, the IT group lost 10% of their staff too. Reluctantly, the corporate mind trust allows the IT team to utilize temporary workers to deliver projects. Of course, there is no commitment for long-term funding so Staff Augmentation becomes the default solution. We have discussed how this perpetuates itself over the long haul and leads to a weakened IT organization. Not only is the health of the IT organization at risk, resource costs are easily 1.5 to 2 times higher than they were before the cost-cutting measures went in! What happened to Economics 101?

The Managed Services Model (MSM) is a better long-term outsourcing solution for the following reasons:

➢ The risk of co-employment issues is lower because the supplier is directly managing the day-to-day work of the third party resources and not the client.

➢ Suppliers are given incentives to deliver specific outcomes instead of tasks and are measured by Service Level Agreements (SLA).

➢ Multiyear contracts spell out SLA's, annual rate card increases, skill levels required and outcomes desired.

➢ The supplier is responsible for training resources when they are replaced (at their expense) in order to deliver the contracted SLA's.

➤ The supplier will install a knowledge management process to capture and document technical and procedural information about an application and how it fits into the technical environment. The knowledge database must remain the property of the client and should be so noted in the contract.

Which model is best to use? Staff Augmentation is a good option for short duration projects that have defined starting and ending dates. Third party resources are brought in, given a specific assignment and then managed by an employee on the client's IT staff. These resources are paid an hourly rate negotiated between the supplier and the firm. MSM is the better choice when a more permanent change in the IT staffing model is desired or required and should be a component of the IT resource strategy. Regardless of which model is used, it is imperative for the client to capture and retain the knowledge needed to manage the organization. Without it, they will not be able to accurately assess whether future work estimates are reasonable. Knowledge loss is the single largest risk with any type of outsourcing. Both staffing models contain certain elements of risk.

Risks with Staff Augmentation:

➤ HR concerns exist regarding co-employment issues.

➤ There is rarely a formal performance contract in place. The client is responsible for all results and work quality.

➤ This model will result in higher cost because of a greater reliance on onshore resources limiting the ability to capture the savings from labor cost arbitrage.

➤ The greatest concern is the loss of internal knowledge. Staff Augmentation resources rarely, if ever, keep documentation and knowledge management protocols current.

Risks with a Managed Services Model:

➤ A solid partnership between the client and supplier is required for this model to be successful. Executive leadership at all levels needs to be linked with their supplier counterparts and meet routinely to discuss quality concerns, service delivery performance, and innovation opportunities.

➤ Over time, a successful partnership will grow larger resulting in more and more business spread across fewer and fewer suppliers. The fear of placing too many eggs in one basket needs to be balanced against the benefits of doing so.

➤ It is common for Managed Services contracts to cover 5-7 years.

By the end of the term, the client's retained staff should only be 10-15% of the total resources. This small team can effectively govern the supplier. However, with the supplier representing 85-90% of the support staff, hitting the "undo" button will be costly, difficult and nearly impossible. So pick a good supplier!!

➤ To minimize the risk of losing knowledge, the supplier must implement a knowledge database to record and store changes to applications in the portfolio. This database can be used to train new resources and maintain the retained team knowledge level.

Deciding on one model over the other on a project-by-project basis is not a good practice. Every IT organization should develop and follow a defined sourcing strategy especially since labor is the single largest controllable cost. A successful sourcing strategy should be developed in partnership with HR, and endorsed and approved by the CEO. When developing the strategic direction for IT, one needs to make sure technology enables business results.

Comparing payment terms:

We have discussed the Staff Augmentation and Managed Services models. Lets spend a moment to compare and contrast the different ways to pay the supplier. The two primary options are either "Time & Materials" or "Fixed Bid" as shown on Table 4-2.

Table 4-2 Comparing Time & Materials to Fixed Bid

Topic	Time & Materials	Fixed Bid
Description	The supplier invoices the client for billable hours and materials	The supplier agrees to deliver a defined outcome for a fixed price
Criteria for Selection	Short duration engagement, non recurring work with a high degree of uncertainty and risk	Routine work with statistics that determine the work effort required of the supplier
Relationship	Client oversees tasks and number of hours used to assure they are getting a good value	Client pays for a specific outcome and should not be concerned with number of hours used to deliver it
Measuring Results	Client needs to audit the work output to validate delivery and invoicing accuracy	The client uses SLA's to assure that the supplier is delivering against agreed to milestones
Combining the two models	- If a supplier has multiple SoW's in play, it is possible to use both - A contractual relationship with metrics is required for Fixed Bid - Time & Materials can be a subordinate SoW to a Fixed Bid contract	

Description:

Under a Time & Materials payment structure, the supplier works on specific tasks for the client. The number of hours required to complete the task is multiplied by the rate for the skill levels used. Any consumable costs such as travel, office supplies, etc. are also added to the total cost charged to the client. The supplier will bill the client regardless of whether or not the desired outcome is achieved unless there is a problem with the quality of the resource.

In a Fixed Bid model, the client is paying for outcomes rather than specific tasks. The outcomes must have SLA's associated with them to assure the client received what they are paying for. Whether the supplier uses 5 resources or 50 to deliver the outcome really doesn't matter, since the price charged to the client does not change. This is a hard concept to grasp, since most of us grew up managing activities (and people) rather than outcomes. As long as SLA's are being met, the number of resources used to deliver them is immaterial in the fixed price model.

I enjoy great food, which is why so many of the examples used in this book are food oriented, and this one is no exception. Reverting back to our steak example, we might want the local restaurant to deliver a cooked rib eye steak to our house. If we paid them on a Time & Materials basis, they would charge for the time it took for someone to go to the store, search the meat case, pick out a steak, cook the steak, drive to your house and present you with the meal. They would also charge you for any materials used to complete the task (namely the steak). In this model, if the first store they visit is out of steaks and a trip to a second store is needed, the actual time spent will increase as will the cost. Also, if the steak at the alternate store costs more, you will pay more for materials too. Under a fixed-bid arrangement, you agree to pay specific price for the restaurant to deliver a steak dinner to your house. If something happens requiring more time to complete the work, you don't care since you will be charged at the agreed price. If the supplier finds a sale on steaks, however, they pocket the savings.

Criteria for selecting one over the other:

Time & Materials is best used for short duration projects or other nonrecurring work with a high level of uncertainty. From the supplier's perspective, they may not be willing to sign up for fixed price for a service until they understand the environment, risks, and work effort required to execute the project. If forced into a fixed-bid arrangement before they have learned the environment, the supplier will price the job to cover that uncertainty.

As a supplier works in the client's environment, they quickly broaden knowledge of the application portfolio and technical environment. This increased knowledge reduces uncertainty and helps them formulate accurate bids on future work. This is good for the client but eventually skews competitive bidding in favor of the incumbent. Low bids from a new supplier will be rejected based on the logic that "the bid is low because they really don't understand the complexity of our environment." Eventually, other vendors will tire of generating bids they have no possibility of winning. This is a bad thing. To avoid it and foster competitive bidding, the client needs to allow ample time for any new supplier to study and learn the environment. A low-risk way to let a new supplier gain that knowledge is to hire them to work on a small development project before committing to a long term agreement.

Relationship:

With Time & Materials, the client needs to oversee the work activities and resources to determine if they are getting a good value for the money. Under a Fixed Bid approach, the client reviews SLA's to assure that the outcomes are being delivered. Fixed Bid means that the client pays a predetermined amount so it really doesn't matter how many resources the supplier uses.

Measuring results:

Fixed Bid is best suited for work that is well documented and associated with a long-term contract. The nature of Fixed Bid requires the use of well-defined SLA's. Without historical data, the supplier may need months to study the environment before they can agree to the SLA targets. In this case it may be better to start the relationship with 6 months of "adjustment" time to allow hidden issues to rise to the surface. The key deliverables, SLA's and price, would be defined before the work begins, however, both parties would have the opportunity to use data collected during the adjustment period to request minor tweaks to the SLA's but not to the price or deliverables. It is a good idea to put min/max parameters on how much the SLA's can change from the original estimates. Together, the client and supplier will review the data and agree on final parameters of the SLA. This approach makes sense for both sides rather than forcing the supplier to agree to a Fixed Bid before they have had a chance to fully understand the environment. The supplier's staffing and cost is driven by the SLA. For example, it takes more resource time (cost) to deliver an SLA of 99% uptime for application X than it would to provide an uptime of 90%. When undisclosed issues with application X are discovered, the supplier has the opportunity, within reason, to work with the client to create realistic expectations and metrics.

Combining payment methods:

The type of pay structure is defined within each Statement of Work (SoW) – either Fixed Bid or Time & Materials but never both. That said if a supplier is engaged in multiple SoW's it is possible that some will be Fixed Bid and others Time & Materials. Since traditional Staff Augmentation suppliers do not manage the work done by their resources, the use of a Fixed Bid payment agreement is not practical. Fixed Bid requires a contractual relationship with performance metrics in place. The default payment method for Staff Augmentation is Time & Materials.

A common practice for suppliers working in a Fixed Bid arrangement is to modify the existing SoW or create a new one for any incremental work the client requests. For example, with a contract for application support work in place, it is common to create a small SoW to address a nonrecurring initiative and then pay for it using Time & Materials. A nonrecurring initiative is defined as a project with a specific end point that was not included in the primary Fixed Bid contract. Time & Materials may also be used to provide support for applications not included in the initial scope of the support contract. Once the application issues settle down, the incremental resource requirements for the new application should result in a permanent scope change to the primary contract. More on contracts later.

The Captive Center vs Outsourcing:

Although Captive Centers are technically not outsourcing they deserve some attention. A Captive Center is nothing more than a company's branch office in a remote country. It is not outsourcing since local resources are hired as employees. However, the work they do is basically the same as what a third party outsourcing partner would provide. Captive Centers, established in low-cost geographies, face issues similar to those experienced with outsourcing. Although many firms are having good success with this model, there are concerns to keep in mind.

➤ Keep expectations realistic. The main reason companies opt for Captive Centers is to bypass the outsourcing partner and capture a larger share of the labor savings. If an IT resource in Bangalore, India is paid $7-10/hr, the outsourcing firm will charge the client $25-30/hr for that individual. The $15-$20 per hour saved is significant and will yield an annual savings of $30,000 to $40,000 per employee. Some "non-labor" costs are more affordable in India too such as medical insurance, employee benefits, worker compensation, and other insurance costs.

➤ Cost management is key to making a Captive Center viable. Although many costs are lower in India, there will be a few that are higher. For example, compared to other countries taxes may be higher in India. Utility costs can also be higher as India imports most of its gas and oil. Not only are electricity costs higher, the service is unreliable and requires the use of generators for emergency backup. The cost of real estate and rents are surprisingly high in major Indian cities as is the price of a good hotel. Captive Centers are often considered as a means to maintain control over proprietary data, intellectual property (IP) and unique technologies. Since these require enhanced security, maintaining secrecy and privacy increases the cost as well.

➤ Captive Centers are an overseas branch of the parent company and require a capable on-site management team to run efficiently. In order to keep attrition rates manageable, Captive Centers must invest in a well-staffed HR team to focus on employee recruitment and training. A large number of employees are required to generate the savings needed to pay for the high fixed and variable overhead costs of a Captive Center. The average staffing is 1,000 employees with a minimum of about 350.

➤ Captive Center resources will need to link daily with the home office in the U.S. to coordinate with project teams and clients for direction and feedback on various IT initiatives. The time zone difference of ten to twelve hours makes this difficult. Thoughtful planning should be done to assure that communication protocols are defined that include standard project team meeting times, how to link with adjunct resources (infrastructure, DBA's, etc) and steps for contact during off-hour or off-cycle times.

➤ Even though the hourly cost of labor is low, the overhead cost to hire, train, develop and retain employees will be comparatively high. The perception many westerners have of the Indian workforce is that there is an endless supply of hard-working, educated resources competing for positions in IT. Unfortunately, recent observations prove this to be exaggerated and top Indian firms need to invest heavily in training new recruits.

o There are over 20,000 colleges and universities in India, yet firms are facing growing difficulties recruiting and filling their demand for high-quality graduates. Some estimate that only 25% of recent graduates or "freshers" as they are called, have the requisite skills to be immediately productive as an IT resource. The rapid growth of Indian IT service providers has resulted in extensive hiring over the past decade placing tremendous pressure on the educational system in India. The highly praised Indian Institutes of Technology, Indian Institutes of Management, and a few other elite schools together enroll less than 25,000 IT oriented students. Although this small sub-set produces top quality graduates, the problem is that their numbers are too limited and are a mere drop in the bucket of national needs. Further contributing to the problem is a lack of qualified faculty. The average pay for "freshers" from the top schools is 10 times the pay of experienced instructors. This creates an incentive for instructors to leave teaching and work in the private sector. If India cannot find a way to improve the quality of its mainstream universities and the more than 20,000 undergraduate colleges affiliated with them, the overall quality of the system cannot rise. This has been a key challenge for years and remains an important factor today.

o   Captive Centers have limited options for the employee's career path relative to what the large competitors can offer. This is one of the primary causes for turnover.

o   The fallout from the supply side is obvious for Captive Centers as they compete with large Indian-based firms such as TCS (Tata), Infosys, Cognizant and Wipro for a limited number of top quality resources. They need to either pay a premium and hire experienced resources or invest heavily in training facilities (or both!). As noted earlier, many Indian based firms have invested in full-functioning training facilities complete with high-tech classrooms, dormitories, swimming pools, bowling, movie theaters, restaurants, shopping and more. It is difficult for a Captive Center to compete with that!!

➤   Increased capacity. Another reason to open a Captive Center is to create additional capacity to meet the anticipated growth in demand for IT services. If there is a 25% increase in demand, the Captive Center can absorb it by using extended shifts and weekends. If the increase is short-lived, perhaps to provide extra support for the implementation of a major application, it might even be met without incremental cost. Longer-term capacity needs can be met using overtime. An outsourcing partner on the other hand will require the client to pay for that 25% increase. Unfortunately, when demand for services drops, the parent company is faced with maintaining a costly overhead structure with diminishing revenue.

Outsourcing and captive operations start out with each having similar driving forces – namely cost reductions and competitive pressures. Table 4-3 provides a side-by-side comparison of strategic and operational factors to help the organization decide which option to choose over the other.

As we discussed earlier, Captive Centers need economies of scale to cover investment and ongoing overhead costs. Therefore, it is not advised to split the organization with some functions using Captive Centers and others the traditional outsourcing relationship.

It may be the case that a Captive Center grows out of an outsourcing partnership. The initial supplier may be utilized to help the client get their center established. Further, Captive Centers may also provide services to others. A few years ago, I had the opportunity to meet with a major U.S. insurance company that was selling IT services from their captive center located in Ireland.

The selected approach will depend on whether the primary driver is short-term cost savings or whether the company has long-term vision for moving its operations offshore to retain control over processes and intellectual property.

Table 4-3 Comparing Captive Centers to Third Party Outsourcing

Captive Centers	Third Party Outsourcing
- Guaranteed resource requirements	- Irregular but ongoing projects or one-off
- Predictability of growth and retraction	- Defined requirements and delivery specifications
- Capable of delivering core services remotely	- Ready to invest in clear specifications
- Dedicated to self managing	- Desire to gain operational flexibility
- Intellectual Property and sensitive business material that need privacy	- No IP or business sensitivity issues
- Direct control of remote team	- Not willing to invest in a long term funding commitment
- Desire to build/retain domain knowledge	- Require access to technology and domain experts
- Strong delivery processes in place	- Comfortable with the external vendor managing non-core work
- Capital investment required	- Minimal infrastructure requirements
- Able to build offshore management teams	- Offshore tasks require a low learning curve for success
- Desire long-term efficiencies for support, software development and other IT functions	
- Regulatory constraints prohibit outsourcing of certain processes	
- Willing to invest in an offshore team of 500 to 1000 resources and maintain it for many years into the future	

Rural Sourcing:

This concept is a slight twist on the Captive Center idea, and is starting to take root in the U.S. as firms establish local call centers onshore. The ideal location is a small town in a rural area with an attractive life-style, a small college, and plentiful low cost resources.

Many of the issues facing Rural Sourcing are similar to those with Captive Centers. It is very possible to create a competitive alternative to outsourcing offshore if the average hourly rate for labor plus overhead costs for facilities and support staff remains competitive to what offshore firms offer. The obvious plus is employing local resources that are native speakers of the U.S. vernacular of English not to mention the good public relations that come from hiring resources in the U.S.

Chapter 5:

Cost vs. Benefit Analysis

This chapter will examine the cash flow of a typical outsourcing project to determine its financial viability and answer the question "Does the project generate enough savings to justify the expense required to implement it?" The cash flow model compares "current" costs to "new" costs as well as any one-time investment required to achieve the savings. Of course there might be other, non-financial reasons to go forward with the project (i.e. lack of internal resources). If so, it is still important to create a cash flow model in order to fully understand the Total Cost of Ownership and budgeting implications.

Cost reduction is a common reason to outsource. It is accomplished by capturing labor arbitrage savings[2] from the use of IT resources in lower cost "offshore" geographies such as India. For illustrative purposes, let us assume that typical rates for a resource working from India may be in the range of 40-50% less than the cost of a fully loaded U.S.-based resource performing similar tasks and with similar experience. At the low end of the range are resources with more common skills such as Java coders. At the high end are resources with advanced skills such as a project managers or application architects. The number of resources used and the length of the agreement will also impact the hourly cost. For analytical calculations, we will use $30/hr for the average hourly rate of an offshore resource. The $30/hr rate is approximately 40% less when compared to the $52/hr cost of a fully loaded IT employee based in the U.S.

A common logic flaw made when outsourcing is to assume all resources on the project will be paid the offshore rate. Depending on the type of work being outsourced and the maturity of the

---

[2] *All rates used in this book are relative and are in no way indicative of what to expect when you negotiate with your supplier. They are simply used to show how to build a cash flow model. Actual rates may be higher or lower than what is represented here and will be the result of negotiations with your supplier.*

relationship, a range of 20%-40% of the supplier's resources will be working in the client's offices here in the U.S. A ratio of 30% onshore is a good starting point with 70% of the team working offshore in India. Typical onshore rates paid for third party resources working in the U.S. are about 40% to 60% higher than the $52/hr cost of that fully loaded U.S.-based IT employee. Skill set, volume discounts and geographic location will affect the price. Due to cost-of-living differences, a San Francisco-based resource will command a higher rate than one with similar skills based in Albuquerque. For illustrative purposes, $75/hr will be used for onshore resources. Many outsourcing firms have centers in Mexico and other Latin American countries. The hourly rates for these resources will be somewhere in between the offshore and onshore rates. Typical rates for "nearshore" are $55/hr. As discussed earlier, the use of nearshore resources is a good way to alleviate time-zone issues. The 40% cost differential between offshore rates and the cost for a U.S. employee is what many CFO's hone in on. However, that old saying "if it seems to good to be true, it probably isn't" applies to outsourcing. Every project will have a mix of onshore and offshore resources. Compare the "blended" hourly rate of the outsourcer's team to see if outsourcing makes sense.

The supplier should be willing to provide full cost transparency in order to understand what drives pricing. The pace of the Indian economy along with exchange rate dynamics has a big impact on costs. Competitive demand for skilled resources will drive salaries up in good times and down in bad. Prepare for a discussion to understand how the supplier's cost structure will impact their Rate Card. Some of the firms I have worked with are very open, others hesitate to provide the foundation assumptions for the fees they charge. I don't believe they are hiding anything; it is just not part of their engagement process.

As important as supplier costs are, it is equally important to understand the cost of incumbent resources, as this is the basis for what you are currently paying. Be sure to include the cost of benefits and annual increases for them as well. Although merit raises are 3%-5% on average, the increasing cost of health care and other employee related expenses are considerable. I use an 8% increase in salaries and benefits.

The cost of office space should be included too. Outsourcing will have a favorable impact on space requirements. For example, let us assume we outsource a team of 200 incumbent resources working in the U.S. and replace them with 30 retained staff plus 60 onshore supplier resources. The remaining 110 will be located offshore in India. This opens up 110 seats allowing the client the opportunity to lower rents and energy costs.

Let's dig into the details behind the labor cost. The outsourcer will usually express resources as Full Time Equivalents (FTE's) rather than in hours. The cost of an FTE is basically their hourly rate times the number of hours they are expected to work in a given year. We use 2000 hrs per year as a rule of thumb. Hourly rates are an important element needed to determine the cost of an FTE. It is also important to note that an FTE may be one individual or comprised of fractions of two or more people (two half-time resources = one FTE). Be sure to discuss fractional resources with your provider.

Example: Project X requires ten FTE's. Seven will be working from India yielding an offshore ratio of 70% and three working onshore. The seven FTE's working in India may actually be ten different individuals with five working on the project full time and four others each charging half of their time to Project X and half to some other project. The total annual cost of these ten FTE's is about $912,000 or $76,000 per month. Table 5-1 displays blended rates and sensitivity to the ratio of onshore to offshore staffing.

Table 5-1 Blended Rate Analysis (using $30/hr offshore & $75/hr onshore)

Relationship Maturity	Percentage working from offshore*	Percentage working onshore	Blended Hourly Rate	Annual Blended Cost per Resource
Before Managed Services	40% working from India	60% working at client site	$58.20 per hour	$116,400 per year
First 12 months	60% working from India	40% working at client site	$49.80 per hour	$99,600 per year
12-24 months	70% working from India	30% working at client site	$45.60 per hour	$91,200 per year
24-36 months	75% working from India	25% working at client site	$43.50 per hour	$87,000 per year
36 months or more	80% working from India	20% working at client site	$41.40 per hour	$82,800 per year

* Offshore resources are paid for 8.8 hrs per day.

Although the math is straightforward, many outsourcing practitioners don't fully understand cost dynamics. Outsourcing firms typically bill offshore resources at 8.8 hours per day for a 5-day week. In Table 5-1, the cost of an offshore resource is $66,000 annually: $30/hr x 8.8 hrs/day x 5 days/wk x 50 wks/yr. Each onshore resource costs $150,000 per year ($75x8x5x50). An onshore resource is paid for 8 hrs per day. I use a 50-week year to allow for 2 weeks of vacation time. If 30% of the resources are working onshore and 70% offshore the simple blended cost is $91,200 (.3x$150,000) + (.7x$66,000). Remember, our rates are for illustration only. Once the actual skill sets and rates are finalized, recast the blended rate using the actual costs from your rate card.

It is equally important to understand the fully loaded cost of incumbent resources in order to accurately perform a cash flow analysis. It is often the case that incumbent resources are a combination of employees and third party resources working in a staff augmentation relationship. The hourly rate for third party resources and hours billed should be clearly visible on the monthly invoice.

An "apples-to-apples" comparison between the fully loaded cost of an employee and that of a third party resource is a challenge to create. There is more to employee costs than the annual base pay. Benefits (insurance, vacation, sick leave, etc.), FICA, and other direct costs associated with having that employee on your payroll need to be factored in. HR and Finance can help determine the cost elements for incumbent resources. IT organizations typically have labor cost data available at the department level with the average being $130,000 per person per year. However, using the department average can be misleading as it includes executive staff too. Generally, commodity work such as analysis, coding and testing is outsourced. Employees at the lower end of the pay scale earning $90,000 to $110,000 per year (fully loaded) perform these tasks. The average cost of those being outsourced is closer to $100,000 per year with an hourly rate of $52/hr.

There are other less obvious costs that should be evaluated as well. I have found it best to convert all work activities into hourly equivalents. IT employees are generally salaried (rather than paid hourly) and their annual pay and benefits are easy to find. However, we also need to know how many hours per year the employee actually spends doing the work that will be outsourced. This is what we are paying the outsourcer to do.

There are two factors that will impact total hours:

One is "uncharged" time and the other is "nonproductive" time. In cases where IT employees are paid a salary, they may be expected to routinely work more than a 40hr week without collecting additional overtime pay – the norm for salaried employees. The "uncharged" time should be factored into the rate comparison. In our example, the average uncharged time is about 5 hours per week or 260 hrs per year. This can vary a lot from company to company and will be discovered during the work activity analysis. Suppliers will consider this time to be part of the workload and will want to charge for it.

On the other side of the equation, salaried employees continue to be paid when they are not performing IT activities. They draw salary while on vacation, at home sick, on paid holidays, and when they attend company-sponsored meetings. On average, employees are paid 440 hrs per year for time away from the job: 15 days of paid vacation; 10 paid holidays; 5 paid sick days and 25 days spent in meetings (assuming only 4 hours per week!). It is important to understand the number of hours employees spend doing non-IT activities as it can represent about 20% of their time. Suppliers will simply assume each employee is 100% engaged in the activities the outsourcer will take over. From the supplier's perspective, each employee spends 2340 hrs (2080+260) per year on the activities being outsourced. However, after subtracting the 440 hrs of nonproductive time, the actual time spent doing work is only 1900 hours per year – about 20% less. This is a pretty big swing and most clients don't realize the impact of this discrepancy. Remember: You are paying the supplier for hours actually spent doing work. Vacation time, meetings, and other nonproductive time have been baked into the rates they are charging.

In the U.S. the average annual salary for an IT employee with four to six years of experience is about $71,500 per year. With 40% benefits, the fully loaded total comes to $100,000. The hourly cost for that IT employee would be $52/hr for the time he is doing work ($100,000 divided by 1900 hrs of work time). The intent of this chapter is to provide the tools and insights needed to perform an independent analysis rather than providing specific costs. When preparing a cash flow model, costs for incumbent resources impacted by the project should be used for analysis.

Another component in the cash flow analysis are the year-over-year cost reductions the supplier builds into the contract. Cost reductions of 5% per year are common and tend to offset contracted rate increases. The supplier basically has two levers to achieve reductions: reducing the total number of resources and/or moving resources from onshore to offshore. In a fixed-bid contract, the client agrees to pay the same amount each year for activities defined to be "in scope." The net of rate increases and productivity savings are also reflected in those payments. With a fixed bid, the supplier wins if they can do the job with fewer resources and lose if they use more. Note: It is reasonable to expect your cost structure could be based on a starting resource ratio of 30% onshore and 70% offshore.

Let's work through a simple cash flow analysis for a small project to outsource application support. The Support Team being considered for outsourcing is made up of two managers each with 9 staff for a total of 20 resources. The actual work to be performed by the outsourcer will be based on everything the 18 are doing. Once the 18 are eliminated, the managers will no longer be needed either. Therefore, the "current costs" in our model will reflect the entire team of 20. Let me be quick to note however, that every outsourcing project needs client resources (10-20% of total team size) for governance of the project. In our example, we will retain two client resources to govern "relationship quality" with the supplier, act as a liaison with business users and evaluate SLA performance. Never let the supplier perform the governance roll! The cost of the two retained staff needed for governance is reflected in the "new costs" in our Cash Flow Analysis.

At an average cost of $100,000/yr, the current costs in Table 5-2 for this work group is $2,000,000. Use whatever annual benefit increase is appropriate for your firm – I used 8% (4% merit and 4% benefits). Over a five-year horizon, the total current cost will be $7,573,202.

Table 5-2 Cash Flow Analysis

Cost Elements	Year 1	Year 2	Year 3	Year 4	Year 5	Total
Current Costs						
18 Employees & 2 Managers	$2,000,000	2,000,000	$2,160,000	$2,332,800	$2,519,424	$7,012,224
8% Merit & Benefit Increases	-	$160,000	$172,000	$186,624	$201,554	$560,978
Total Current Costs	$2,000,000	$2,160,000	$2,332,800	$2,519,424	$2,720,978	$7,573,202
New Costs						
18 FTE's (70% offshore)	$1,641,600	$1,641,600	$1,690,848	$1,741,573	$1,793,821	$5,226,242
3% Annual Rate Increase	-	$49,248	$50,725	$52,247	$53,815	$156,787
5% Annual Cost Reduction	-	-$82,080	-$84,542	-$87,079	-$89,691	-$261,312
Data Connectivity to India	$50,000	$50,000	$50,000	$50,000	$50,000	$150,000
2 Retained Staff – Governance	$200,000	$200,000	$216,000	$233,280	$251,942	$701,222
8% Merit & Benefit Increases	-	$16,000	$17,280	$18,662	$20,155	$56,098
Total New Cost	$1,891,600	$1,874,768	$1,940,311	$2,008,684	$2,080,042	$6,029,037
Net Savings	$108,400	$285,232	$392,489	$510,740	$640,936	$1,544,164
One Time Costs						
Travel	$50,000					
Severance for Impacted Staff	$450,000					
Knowledge Transfer Transition Cost	$325,000					
Data Transmission Set-up	$75,000					
Total One-Time Investment	$900,000					
IRR (Using MS Excel) =	24.5%					

Keep in mind that the 18 performing support tasks, not their managers, are doing the actual work to be outsourced. It is important to note this distinction. I have seen plans where the entire work group (20 in our example) would be replaced by 20 supplier resources. With the two governance staff the total work group grows to 22 instead of 20 – an overall increase of two people!! This is why we have discussed the importance of defining the work elements to be outsourced (not people) and differentiating it from work that will NOT be outsourced.

The "New Costs" in the cash flow model are based on the number of supplier resources required to perform the actual support tasks. The work element analysis provided an accurate picture of the work that can be outsourced and that which can't; as well as the number of FTE's associated with it. The work analysis may also point out activities that are no longer needed or should be performed by another team (i.e. application development). The initial mix assumes that 70% of the work will be performed offshore and 30% onshore. From Table 5-1, the average annual cost for these resources will be $91,200 for a total of $1,641,600 in the first year. We will include supplier rate increases of 3% per year along with a 5% annual cost reduction per year.

There are a number of "One Time" costs noted at the bottom of Table 5.2 to include in the analysis as well. Think of these as the "upfront investments" needed to capture the savings the project will generate. A cash flow model uses the savings to create a return on the up-front, one-time investments noted below:

➢ Travel for retained staff to visit India as well as supplier resources to come to the U.S. to learn the applications during Knowledge Transfer.

➢ Severance for staff if/when their employment is terminated.

➢ Knowledge Transfer - where the new staff learns the job from the current staff. (Incumbent and new staff are bot paid in KT).

➢ Technology set-up costs to link offshore and onshore.

➢ Consulting costs to help plan and manage the overall transition.

The total one-time costs in our model total $900,000. The savings associated with this investment will yield an Internal Rate of Return (IRR) of 24.5%, which is very respectable. The finance team member on the Working Team should perform the cash flow analysis and use internal evaluation tools and hurdle rates.

We have based the analysis on an often-used rule of thumb: new team (client + supplier) will be the same number of resources as old team. For the sake of simplicity, the 1:1 replacement model is what the example in Table 5-2 is based on. Although it is okay for a quick analysis, it may leave a lot of money on the table.

As noted earlier, offshore resources are typically paid for 8.8 hours per day. It is therefore reasonable for the client to expect that an offshore resource will actually work 8.8 hours per day or 2200 hrs per year (8.8 times 5 times 50). In our example, it takes 34,200 hours per year to complete the work of the 18 internal employees (18 times 1900hrs). Five supplier resources would be needed for the 30% of the work (10,000 hrs) to be done onshore. Using a 1:1 replacement model would mean 13 resources would be required to complete the 70% of the work (24,200 hrs) being performed offshore. However, because offshore resources work 8.8 hrs/day, only 11 would be needed (24,200/2200) instead of 13.

When we recast the cash flow model reflecting a staffing plan of 11 offshore and 5 onshore the IRR increases from 24.5% to 43% and yields additional project savings of $517,000 – a 33% improvement in overall savings. I can't emphasize enough how important it is for the client to do their homework and identify the actual work the outsourcer will perform; the number of hours required to perform it; and the number of hours each resource contributes.

Note, the cash flow model we have created is a realistic estimate and is good for planning. During the supplier selection process, use the cost estimates provided by the supplier. Work with the supplier to fully understand all of the detailed assumptions provided in their proposal.

Miscellaneous thoughts on cash flow analysis:

➢ Using the 8.8 hr workday for offshore resources improves the IRR to 43%.

➢ The onshore to offshore mix ratio has a big impact on costs. If the blended rate for third party resources is based on an offshore ratio of 60% vs 70%, the IRR drops to 8%.

➢ The model is sensitive to the number of years it is projected into the future. In our example we used 5 years. Good projects with modest returns may not meet the hurdle rate if the number of years being modeled is too short. If we only projected out 3 years, the IRR in our initial model drops from a plus 25% to a negative 5%. I would suggest extending the model a minimum of five years or longer if the contract is longer than five years.

The Working Team should build a simple spreadsheet to model cash flows and test the sensitivity of various data parameters on overall results. The knowledge gained from the modeling process will help identify critical points to discuss when negotiating with the supplier. It will also help create important talking points for the "go, no go" discussion with the Executive Team.

Make sure to include the cost reductions the supplier has agreed to deliver. There is a perception that all Indian IT resources work harder and are better educated than U.S. resources enabling them to perform superhuman feats. This perception is simply not true. Remember, it is called "Managed Services" not "Magic Services!" The advantage held by Indian outsourcing firms is they have well developed processes designed to identify failure points in the code and correct them. They invest time training their resources on how to optimize results by using these processes. As improvements are made, incident rates decrease and the work can be done with fewer resources – this is one way to achieve productivity targets.

Don't forget to include the governance resources in the cash flow model. With outsourcing, new skills are needed to manage the supplier and build the relationship. In our example we assigned two employees for governance work. We added their salaries, along with annual merit and benefit increases of 8% to the model. In general, the IT employee morphs from "doing" to "reviewing" the work of the outsourcer. Specific duties associated with governance will be covered at length later in this book.

There will be other recurring costs that should be included in the model as well. I have added $50,000 per year for the cost of data connectivity between the U.S. and offshore. Annual travel costs to meet with the offshore teams should be included in the "new costs" as well.

➢ If the sum total of new costs is lower than current costs, the difference will be the savings generated by the project.

There are many costs to be aware of when building the model. Some may pertain to your project; others may not.

A recap of the cash flow model follows:

1) Identify the Current Cost of incumbent resources.

   a) Activities roles & resources of the group being outsourced.

   b) Determine work activities to be performed by the supplier.

   c) Fully loaded cost of the resources doing this work.

   d) Annual increases such as merit raises, fringe increases, etc.

   e) Determine both the uncharged and non-productive hours of the incumbent staff.

   f) Calculate total current costs, plus increases, year-by-year for at least 5 years.

2) Determine New Costs.

   a) Based on the work activity analysis, determine the number and cost of third party resources needed to perform the work.

   b) Start with 70% of the work done offshore and 30% onshore.

   c) Offshore resources paid 8.8 hrs per day; onshore 8 hrs per day. Be sure to include scheduled rate increases.

   d) Include annual rate increases per the agreement.

   e) Include annual cost reductions per the agreement.

    f)   Include the number and cost of the retained governance team.

    g)   Calculate projected new labor costs over the five-year horizon.

3)   Identify other ongoing incremental costs for the model.

    a)   Identify annual infrastructure charges and investments.

    b)   Incremental hardware and software requirements.

    c)   Data transmission, connectivity and data security audits.

    d)   Workstation imaging, maintenance, etc.

    e)   Travel for client staff to visit offshore teams.

4)   Projected Savings – the difference between current and new costs.

5)   Include one-time costs required to achieve the savings.

    a)   Knowledge Transition costs from incumbent resources to the new supplier.

    b)   Supplier and client travel between onsite and offshore facilities.

    c)   Contract penalties (if any) owed to incumbent supplier.

    d)   Right to Use[3] (RTU) fees that may need to be paid to software vendors to allow third party resources to use their software.

    e)   Costs for technical connectivity and security offshore.

    f)   Employee severance and/or layoff costs.

    g)   Temporary transition management fees.

    h)   Anticipated changes to foreign currency exchange rates.

The cash flow model described above offers a quick, low-cost methodology to perform a preliminary estimate of project viability. Use it before engaging suppliers in a formal Request for Proposal (RFP). Although the rates provided in the example are a good proxy, use the most representative rates you have available when you perform the preliminary assessment. The supplier feedback during RFP will provide rates and costs that are more specific to your project. Keep the cash flow model updated as assumptions

---

[3] *The organization works with legal and finance to contact all software vendors to request permission for the third party supplier to use software licensed to the client.*

change during discovery. It will provide the Working Team with a tool to keep tabs on financial viability during the course of the project. The cash flow model is important work and needs to be done properly to assure that the outsourcing engagement is built on solid ground.

The work activity analysis will give the Working Team the chance to learn a great deal about the dynamic processes within the functional area that is targeted for outsourcing. This knowledge will generate huge returns during the course of the project.

When the Working Team is comfortable with the preliminary analysis and the stream of benefits, it is time to review the recommendation with the Executive Steering Team. Once the EST gives the green light to go forward, it is time to invite 3-4 top tier suppliers to create proposals and bid on your project. The formal Request for Proposal   process is explained in Chapter 11. Remember, as you go through supplier negotiations, every cost on the page is open for negotiation.

For many outsourcing projects, unplanned incremental startup costs present a problem from a cost-center budgeting perspective. This is even true for projects with a high IRR. In our example, the total upfront "investment" required to realize the savings is $900,000. Making matters worse, the $900,000 for our project is "expense" rather than "capital" meaning the expenditure will hit the books in the year it is spent resulting in a potentially going over budget. Someone needs to approve going over budget by this amount. In tight times, even with predictable savings in future years, it is difficult for firms to come up with the funds to cover the onetime costs. This is a problem the EST should resolve well before launching the project. Provisions should be made ahead of time to set aside a specific amount to fund startup expenses. Don't proceed without secured funds. Here are some suggestions to offset or at least minimize the budget impact:

➢ Finish the outsourcing work as early in the year as possible allowing the savings stream to flow earlier thereby creating budget favorability that can be used to offset some or all of the incremental startup costs.

➢ The use of "Restructuring Reserves" is common where a specific amount of money is set aside to cover incremental costs associated with restructuring across the enterprise.

➢ Another approach is for the supplier to amortize their KT costs over the life of the project. I am not an advocate approach as this opens a can of worms from a revenue recognition perspective. Review this option with finance and legal.

This concludes Section 1 – Building the Business Case for Outsourcing. Future Sections will focus on Organizational Readiness; Supplier Selection; Implementation and Ongoing Management.

Section 2:

Organizational Readiness

Section 1 focused on building the business case for outsourcing. We concluded it by working through a cash flow methodology designed to quickly show whether a particular outsourcing project, or perhaps outsourcing as a whole, is a direction the organization might benefit from.

Assuming the pursuit of outsourcing is attractive, the next step is to determine if the organization is ready for the changes that outsourcing will require. There are a number of concepts to consider as outsourcing significantly impacts how the organization functions; the culture; and how it delivers value. I strongly caution against going forward with any outsourcing initiative until solid commitment has been made by top management to enact the changes required for success.

Section 2 will begin by reviewing the readiness assessment and then evaluate potential organizational design changes. Next we will discuss key elements of a successful plan and how to manage impacted staff. We will conclude Section 2 by examining a new concept known as Supplier Relationship Management (SRM).

Areas of Focus: *"The Inside of Outsourcing"*

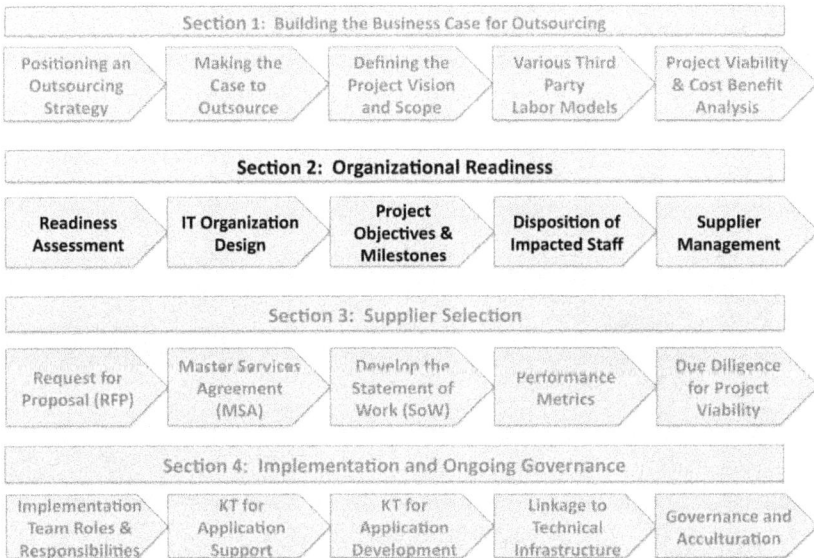

Section 1: Building the Business Case for Outsourcing				
Positioning an Outsourcing Strategy	Making the Case to Outsource	Defining the Project Vision and Scope	Various Third Party Labor Models	Project Viability & Cost Benefit Analysis

Section 2: Organizational Readiness				
Readiness Assessment	IT Organization Design	Project Objectives & Milestones	Disposition of Impacted Staff	Supplier Management

Section 3: Supplier Selection				
Request for Proposal (RFP)	Master Services Agreement (MSA)	Develop the Statement of Work (SoW)	Performance Metrics	Due Diligence for Project Viability

Section 4: Implementation and Ongoing Governance				
Implementation Team Roles & Responsibilities	KT for Application Support	KT for Application Development	Linkage to Technical Infrastructure	Governance and Acculturation

The *Inside* of *Outsourcing*

Chapter 6:

Organizational Readiness Assessment

The Readiness Assessment described in Table 6-1 will be the basis for our discussion in this chapter. These topics form the backbone of the "Change Management" process that is critical for success. We will examine each of these points in detail.

Table 6-1 Outsourcing Readiness Assessment

Strategic Concerns	Tactical Focus
Business Partners	- Business Partner satisfaction and how they will benefit - Changes that will impact the Business Partner
IT Organization	- Mix of internal/external resources - Determine which jobs to outsource - Changes to organization design and new roles needed - Changes to the culture - Governance structure needed
New Processes Needed	- Help Desk and Call Management Center - Performance Reporting - Knowledge Management - Supplier Relationship Management
Skills Readiness for IT Staff	- Review new skills and managerial capability required for the Managed Services Model
Financial Readiness	- Rigorous business case will assure appropriate cash flow and business benefits are attainable
Technology Readiness	- Mature technology control and consistent standards are needed to govern and guide the partnership
Application Readiness	- Documentation and Knowledge Transfer plan is in place and resources are identified to manage transition
Infrastructure Readiness	- Important to connect the new supplier to established infrastructure processes

Business Partner Readiness:

Embarking on an outsourcing project will not only change the dynamics within the IT organization but more importantly, it will have a significant impact on the business users as well. Nothing creates problems faster than disparate expectations. I have long believed that the best way to achieve business user satisfaction is by following this simple thought:

Manage change by managing expectations! Although this is easy to say, it can be quite difficult and exhausting to accomplish.

Elements needed to build a stakeholder partnership include:

➢ Involve the Business Partners in major decisions, project plans and milestones.

➢ Keep them aware and involved in the assumptions used to develop the business case.

➢ Make sure they know what new services will be offered and which existing services may be eliminated and/or changed.

➢ Explain to your Business Partners how existing interactions with the IT organization will change and how their requests will be managed in the future.

➢ Ask them to participate in Service Level discussions. Request their input and sign-off on SLA's and also their help and feedback as you define incident severity levels.

➢ Conduct post-install discussions with them to get their feedback on how the new model is working from their perspective.

➢ Make sure to include them in annual supplier reviews to assure alignment and acceptance of service delivery objectives.

➢ Involve Business Partners in ongoing discussions with the supplier to generate productivity and innovation ideas.

IT Organization Readiness:

Outsourcing, especially to a Managed Services Model, requires certain organizational structures and concepts in place to facilitate a successful transition. Outsourcing doesn't manage itself. A permanent management team is needed that will directly manage the transition and optimize knowledge transition to the supplier.

After implementation, Working Team members transition into the role of supplier governance. This retained team is often staffed with members of the current incumbent staff that didn't get rationalized – hence the term "retained." The number of retained staff is typically 10-30% of the entire team size. For example, if the team for a support project is 200 resources, 20-30 employees are needed to form the governance team. Pick wisely. The retained team makes it happen!

Note: Once the project is finished, it is important that the implementation team becomes the governance team. This will help ensure sure there is skin in the game and no "short cuts" are taken. Problems occur when companies select one team for implementation and another for ongoing governance.

Existing roles, such as Business Technology Engagement (liaison between IT and the business) will be relied upon heavily as will resources from HR, Legal, Procurement and Finance. Although time-consuming, they should be able to provide the support needed without adding permanent incremental positions.

With outsourcing come job eliminations. Work with HR and the compensation team to assure that all jobs (especially those in scope) have updated descriptions. Even without outsourcing, it is a good practice for each employee to update (and/or create) a skills inventory for himself or herself.

As firms rely more and more on collaborative relationships with third party providers, Supplier Relationship Management (SRM) has emerged as a new role with a critical set of processes and dedicated staff to manage. Chapter 10 is dedicated to SRM.

Determine if the organization is structured to create the greatest opportunity for a successful transition to the Managed Services Model. It may be appropriate to restructure in order to provide focus and energy on specific areas that will spearhead change.

Outsourcing will drive a culture change across the organization.

➢ The shift from "command & control" to "managing SLA's" sounds simple but it takes time and constant training to assure it becomes engrained in the retained organization. Employees need to avoid the temptation to "step in to fix issues." Minimize value leakage by letting the supplier do it.

➢ Working with an offshore team that is 10-12 time zones ahead requires a different way to engage with the supplier's resources. For example, there is no "good" time for a global conference call. Someone will be required either stay up very, very late or start very very early.

➢ A cultural workshop designed to help understand the gestures, customs, personality nuances and Indian holidays (different than in U.S.) can be extremely helpful.

Governance takes place on a couple of different levels. The Supplier Relationship Management group governs in a "top-to-top" mode assuring that strategic directions of the supplier mesh with those of the firm. Supplier Governance, on the other hand, takes place at the worker level. Here it is important to assure that resources are properly trained; services are delivered as described in the contract; knowledge is captured in Knowledge Management databases and the 3-legged stool of User-Supplier-IT is balanced and moving in a positive direction. Successful governance results in a high level of user satisfaction and process quality.

IT Process Readiness:

The following IT processes with the exception of Supplier Relationship Management are neither unique to, nor required for, outsourcing. They are common-sense approaches that will enhance the effectiveness of the IT department and at the same time, enable successful outsourcing initiatives. Business Requirements Analysis formalizes conversations between IT and the BP's. Many clients have a point of view on the specific "software packages" they want installed to help manage their business. However, more important than the software package itself are the business problems facing the client and the functionality or "outcomes" required to solve them. Although application "X" may be desired, application "Y" may provide the same functionality but is a better fit for the infrastructure and much less costly to install and maintain.

➢ The goal of initial discussions should focus on the problem that needs to be solved and savings associated with solving it rather the specific software solution.

➢ Once the problem has been properly described, various alternatives to solve it may be considered along with a cost estimate to implement each alternative.

➢ The Business must be willing to provide leadership, people, and user training. They need to lead "business" projects resulting in new tools that will change how users do their jobs. IT will manage the technical components of the project but needs the business to take the overall lead.

Intake and Demand Management formalizes the list of projects and initiatives the IT department must deliver.

Traditionally, the functional silos of yesterday's IT structure left each team to their own devices to gather business needs, request funding, and manage results. These often lead to conflicting (and sometimes redundant) efforts with nobody in control of the broad IT budget.

The foundation of the Demand Management concept is a Project Portfolio; a complete listing of all IT projects along with estimated costs and annual benefits. Projects are either non-discretionary or discretionary.

➢ Non-discretionary spending is for application support and the funds needed to keep the infrastructure and hardware operational. It essentially means, "keep the lights on."

➢ Discretionary spending can be further broken down into "Strategic" vs "Investment" projects.

➢ Strategic initiatives are typically activities that support other projects but may not have a recognizable financial benefit on their own. An example is the creation of common data structures or Master Files. They are required to efficiently manage common data across the multi-application suite.

➢ Investment projects generate a measurable benefit: The Internal Rate of Return (IRR), absolute dollars generated or both.

➢ A new role, Portfolio Manager, creates and manages a prioritized list of projects ranging from non-discretionary to strategic investments. The IRR and total savings may influence ranking may not address the full picture. A project with a 50% IRR may deliver $1 Million in savings, yet one a 25% IRR may deliver $5 Million – which should be ranked higher? The Portfolio Manager must understand the broad needs of the organization as well as the level of risk associated with each project. The $5 Million savings is more attractive, but the risk to deliver it might be unacceptable.

Call Management Systems are used to formally link business users to the Help Desk. When incidents occur, or a request for service exists, the user has a common Help Desk number to call. The Help Desk can perform many routine automated tasks such as "password resets." Scripting is the backbone of the Help Desk and must be done well. The Help Desk opens a ticket and routes the ticket, based on the scripting, to the correct queue for resolution by the application Support Team.

The "ticket" is an electronic identifier containing a wealth of information including the time of initial contact; incident description; the time the ticket was acknowledged by the Support Team; the time the incident was resolved; and a description of how the incident was resolved.

➢ The data collected from the "ticket" become the elements used to measure Service Level Agreements (SLA's) with the supplier.

➢ Without a high functioning Help Desk and associated Call Management software, the ability to manage SLA's and generate any sort of "failure point" analysis is simply not possible. Failure point analysis identifies recurring issues leading to enhancement plans to address them.

> The Help Desk also route requests for service. A user may ask for a certain software package for their workstation. The ticket is routed to the correct person who approves and manages the installation and then bills accordingly.

Performance reporting evaluates how well the supplier partnership is doing against predefined SLA's in the case of support work and milestones in the case of development projects.

We have touched on the need for a Help Desk with Call Management tools for managing support SLA's.

Development projects require that milestones be created before the project starts. Believe it or not, this is a major change to the way many SoW's and/or project plans were written. Managed Services requires that SoW's be written with outcomes in mind. The old way of writing a SoW was "send me 10 Java coders for 3 months." The new way of writing a SoW more complex and is based on describing the business outcomes along with an agreed timetable for achieving them.

Knowledge Management and Knowledge Retention, or more precisely the loss of it, is by far the single greatest risk associated with outsourcing. Knowledge Management is a collection of processes that govern the creation, dissemination, and utilization of knowledge. Firms must take action to prevent the loss of internal resource knowledge about the detailed technical workings of their IT environment.

Supplier Relationship Management (SRM) will manage the overall use of third party IT service providers to assure that:

> Sourcing processes are consistent across all suppliers and will conform to organizational needs instead of the organization conforming to the needs of each individual supplier.

> Fair and ethical practices occur on both sides of the relationship.

> There is a common point for legal to work with assuring that supplier contracts contain common terms and conditions that enhance the outsourcing success for the client.

> There is a common point for IT project managers to share current and future needs for third party services to better manage the flow of resource needs and incorporate them into supplier discussions as appropriate.

- ➢ Rate cards and pricing terms reflect the best interests of the company. A common example would be volume based discount incentives.

- ➢ All third party work is bid on a competitive basis.

- ➢ There is a common escalation process when issues with a supplier need to be addressed at a higher level.

- ➢ SoW's and SLA's are worded uniformly across all vendors.

- ➢ Payment terms and payment processes are consistent across the pool of technology suppliers.

Skills Readiness – Management and Staff:

As the migration to MSM occurs, job skills that were effective in the old "command & control" organization change to those that foster "relationship management." It is important to point out that not everyone on the old team will make the transition – some will not want to and some may not be able to. Before finalizing the retained team, a conversation with each candidate should take place to understand his or her goals and career aspirations. New skills will be required for a successful transition. In order to manage outcomes rather than tasks, the governance staff needs to focus on strategy, business engagement, requirements gathering, and, most importantly, building a partnership with the supplier.

I have seen successful partnerships and suffered through horrible ones. The common denominator for success was when the client didn't hold the supplier to the letter of the law on the contract nor did the supplier hold the client to the letter. The lens and perspective through which the supplier is viewed needs to change from that of vendor to true business partner who could help the firm achieve results. Negotiating, discussing and working through issues to arrive at a "best for all" solution is the foundation for partnership. The client will win some and the supplier will win some. Realize that everything cannot possibly be documented, negotiated & spelled out in a contract. The common denominator in those horrible relationships escalated when one side or the other whipped out the contract trying to penalize or hold the other side to the letter of the law all the time, every time. As a direct result those relationships became hostile and ineffective.

Once the transformation to MSM is complete, the supplier will comprise 70-80% of the total resources base. The supplier manages the day-to-day activities of their resources and the retained client staff from the old IT organization will need to "manage the partnership" instead of the direct resources.

New skills required for managing the partnership include:

➢ Negotiating instead of using the contract as a club.

➢ Defining the outcomes to be delivered by a given project.

➢ Planning for resource coverage to assure that required supplier knowledge and skills are in place when needed.

➢ A working knowledge of the contract to make sure the client can optimize value by avoiding the most common mistake of "doing the vendors job for them."

➢ Keeping morale high and identifying resource burnout.

"Managing against outcomes and SLA's versus tasks and activities" is an often-used concept that bears a bit of explanation. The traditional interaction with a technical services supplier is to describe the specific activity to be done – for example write 10,000 lines of code that follow predefined specs.

➢ In MSM, the client needs to describe the outcome the code will achieve. For example: create a system to pay invoices and match them against approved Purchase Orders (PO's).

➢ For this to work properly, the supplier needs to work closely with the IT project leads and the business clients in order to understand the business processes and learn the current application architecture.

Higher order in-house skills are required. The role of Application Architect is vital for development engagements.

➢ The Application Architect will need to be strategic and manage all phases of both business and supplier engagement.

➢ The Application Architect must know about current and future business processes as well as new requirements.

➢ They also need to be able to manage the various phases of project development from initial design through testing.

➢ Most importantly, they need to understand current and future application architecture and how it fits into the existing technical infrastructure.

The underlying technical platform for each organization is different even though they may be using the same tools. This makes it difficult to hire Application Architects off the street. Top resources will be groomed over a couple of years. Tier 1 suppliers can develop the architect role, however, given the critical nature of the role and close working relationship with the business partners, my suggestion is to develop that expertise in-house.

Financial Readiness:

In Section 1 we developed a business case and financial model for outsourcing. We covered various aspects of outsourcing to ensure a solid understanding of value delivery. We went into great detail on cost drivers and developed a simple cash flow model that included onetime costs. There is no reason to repeat that section here.

Technology Management Readiness: Outsourcing, whether for application support or development, requires consistent technology standards and a reliable environment in which to host applications.

Technical standards for hardware drive efficiency both in how they operate and how they interact with each other. Consistently applied configuration protocols and version level standards for servers, as an example, make trouble shooting much easier for those responsible for keeping the lights on. Consistent standards drive how new applications must be written to fit into the environment.

Architecture standards include technical platforms and also define how data flows from application to application residing on those platforms. Organizations often grow through acquisition resulting in complex suites of systems that are somehow connected together. This explains what appears to be a "mish-mash" of systems and servers in large companies. Establishing a future-state vision for architecture gives all supplier partners a road map to follow in hopes of driving out complexity. Unfortunately, the cost of getting to a more efficient future often outweighs the hard dollar benefits of getting there. Then again, when personal computers were first being offered in the mid 1980's, it was hard to cost justify one. Look at what the world of personal computing has evolved to today! Defining and funding the future-state architecture vision should be on the top of the "to do" list for every CIO.

Application Readiness: Each application should be in a stable state and are experiencing major changes. It is a good idea to postpone these applications and transition them a later date.

Documentation is boring, time-consuming to maintain and is rarely ever looked at after the application is installed. These are all reasons why documentation is in such an abysmal state in most IT organizations today. One of the benefits outsourcing to a top tier firm brings is improved documentation. In the U.S. there is a strong reliance on people, or in many cases a specific person, with tribal knowledge of how the application works. Outsourcing firms are not people-dependent but rely on well-designed processes instead. The foundation for process management is application documentation that is easily accessible and maintainable when modifications are made to the application. During Knowledge Transfer the supplier will fill in missing documentation and will use it to train junior replacements. The Governance team should check to assure documentation is being maintained properly with all code and environmental changes.

One way the organization can minimize the length of time (as well as the cost) for Knowledge Transfer is to create quality documentation before engaging a Managed Services provider. Without it transition time is extended.

Infrastructure Readiness:

Infrastructure is a combination of hardware elements and the programs that run them such as telephony, servers, databases, middleware, etc. Infrastructure management may be outsourced and when it is, there should be Operating Level Agreements (OLA's) in place between the supplier of infrastructure services and the other IT services vendors.

OLA's will define how multiple vendors will work together for the good of the enterprise. When an application stops operating as designed the fault may lay within the application or it may be the result of a server issue. Resolution often requires that vendors work directly together with OLA's defining the exchange. For example, both vendors should work under similar response parameters for the severity of the incident - Sev 1 issues should be responded to in 15 minutes. It sounds simple, but getting two vendors to agree on common metrics is challenging. Getting them to agree to act as one, without significant penalties is close to impossible.

> I have seen examples where my staff had to play mediator between the infrastructure vendor, the support vendor and the software vendor because none of the three wanted to work with the others. This is simply not acceptable behavior and should be pointed out as such in the contracts as well as during routine operational reviews with the suppliers.

The degree and completeness of infrastructure documentation will determine the difficulty of transition to an outsourcing partner. The use of a Change Management Data Base (CMDB) to track code modifications, configuration settings on the servers, OS Levels, etc. will be an invaluable tool to optimize the outsourcing experience.

An Outsourcing Readiness Assessment provides a concise overview of the many preparations and change management actions needed to assure successful outsourcing. Of course each organization is different. Each must be evaluated independently in order to focus time and resources on the most pressing changes required.

Chapter 7:

IT Organizational Design

Will outsourcing require a change to the IT organization structure? As we saw in Chapter 6, the Organizational Readiness review may result in the need to adapt or change some internal processes. At the same time, there may be other dynamics in play that should be considered. Allow me to share some observations with you.

Ask yourself if the organizational change will actually improve operational efficiency or will it simply shake up the status quo. Sometimes the actions associated with enacting change have more of an impact than the resulting change itself. There is an observed phenomenon known as the Hawthorne Effect (coined in 1950 by Henry Landsberger) in which short-lived worker productivity improvements were observed regardless of changes made to the work environment.

> ➢ The studies were commissioned at the Hawthorne Works (outside of Chicago) in the mid 1920's to learn if lighting levels affected worker productivity. An experiment was designed to measure worker output under various lighting levels. Worker productivity seemed to improve regardless of lighting intensity and slumped when the study was concluded. It was suggested that the productivity gains occurred because workers were more motivated by the interest being shown in them rather than by changes to the actual lighting level.

We have been working in offices and factories since the late 1800's and one would think that firms should have stumbled on the optimal organization design by now. However, to stay competitive in a world of dynamic change, organizations need to respond to environmental changes happening around them. Technological innovations continually change the paradigm within which a business operates. For example, heavy reliance on computer technology started in the 1950's and the growth in this area has been nothing short of phenomenal. I just paid for my cup of coffee with my iPhone! The technology supporting this was not even thought of let alone possible thirty years ago. One element of organizational evolution is the advent of the IT Department itself (which didn't exist prior to the late 1950's) providing expertise to assimilate technologies into the workplace.

I believe organizations change their structures way too often – every 2 or 3 years in some cases. Sadly, the intent of the change is often to reward or punish someone rather than to actually improve operational efficiency and effectiveness.

> For example: Mary has done a great job as VP of Accounting so lets reward her by changing boundaries and have the Procurement function report into her as well. That way, we can eliminate Sam's position as VP of Procurement. Sam's performance is marginal and he isn't well liked anyway. Restructuring with the goal of job elimination is becoming the approach *du jour* rather than creating performance improvement plans. It is much easier to tell Sam that due to efficiency targets his job has been eliminated, than it is to tell him he is being let go because of poor performance.

It is true that outsourcing has a profound impact on organizational dynamics. The key elements associated with outsourcing include working with resources half-a-world away; coordinating work with multiple suppliers; and learning to manage outcomes instead of worker inputs. As we learned from the readiness assessment in the prior chapter, there are a number of processes that need to be in place to enable an effective transition to a Managed Services Model (MSM). MSM, as its name implies, engages a third party supplier to perform work and relies on a strong partnership with that supplier to achieve multiple objectives.

The outsourcing vision will determine the level and sequence for making organizational changes. Organization changes associated with large outsourcing initiatives include: Centralizing Application Support, creating Demand Management protocols, implementing a Call Management Center as well as a formal Supplier Relationship Management (SRM) function.

We should certainly consider adding new processes where required but at the same time we need to take a closer look at redesigning how existing tasks are performed. Over time, as the IT organization grows, its priorities and skill sets may change to reflect increased complexity. At some point along this continuum there may well emerge a need to change the organization structure.

Rationale for Centralizing the Application Support Function:

In a small "start up", technical capabilities grow in response to functional needs. The overall application landscape is generally quite limited and is managed by a scattering of small independent IT teams each representing a specific function such as supply chain, inventory, order management, etc. The emerging IT shop cannot afford, nor does it require, two separate teams for Support and Development. Having one team performing both roles promotes speed and flexibility in product development and deployment in small IT shops. Sometimes, the overall suite can be developed on one (or a few) application technology stacks; for example, customized power builder, visual basic or Java applications. In such cases, using Enterprise Resource Planning (ERP)[4] solutions such as SAP, ORACLE and Sage with their associated license, infrastructure and heavy implementation costs does not make economic sense.

As the company grows, the demand for more enterprise tools drives them to seek and engage third party help via the Application Service Provider (ASP) route. ASP's have been around for a number of years and some offer to host their applications for the client charging them on a "pay-by-the-drink" basis (i.e. per transaction), which is fine as long as the transaction volume remains low. In many ways, the buzz associated with cloud computing mimics this.

As volume and complexity in business operations grow, the company evolves into a larger enterprise. It is at this point where we often see the formation of a centralized IT organization. Now centralized, the IT organization can capture scale efficiencies and develop strategic plans that affect the entire business. These strategic plans are often based on integrating existing legacy applications to an ERP solution. The new "ERP + legacy" solution results in more data passing through a deeper set of functional capabilities within the vertical domains of supply chain, inventory, order management, etc. At the same time, more data than ever before is being passed and transformed _across_ the enterprise linking multiple applications through interfaced layers of synchronized batch schedules or asynchronous real-time services.

The simultaneous occurrence of deeper functionality within each vertical and more data flowing horizontally across the verticals drives two interesting trends. One is in the difference in skill sets required and the other is prioritizing work assignments. On one hand, more specialized analysts and technologists are required that are well versed in the deep functional and technical knowledge. On

---

[4] *ERP solutions make data available to many applications/functions across the enterprise resulting in improved operational vision and decision-making. The three largest global ERP vendors are SAP, ORACLE and Sage. Sage is the largest supplier to small firms while SAP and ORACLE share the top spot for large firms.*

the other hand, the horizontal linkage between applications requires generalists that understand cross-functional data transformation, information roll-up and system integration. The other trend that emerges relates to how the different teams assign priorities. The vertical team is usually more attuned to the business objectives relevant to their area. Typically this is associated with new development work with the team driven by project deliverables, timelines and budget. The horizontal team is more attuned to how data is transacted and flows as well as how the systems communicate and work together across the enterprise. In the model where support and development are in the same group, the development agenda drives the work priorities for all resources, including those of the Support Team.

A savvy IT leadership team knows that a strong development focus is needed to successfully deliver projects that increase business value. At the same time, a strong focus is also needed on application support in order to provide operational stability and service. Development and Support Teams should not be oppositional requiring a traffic cop to mediate – they need to complement each other. They can't be solved with one goal, just as you cannot draw a square circle. Therefore, it makes sense to consider restructuring the IT organization creating a centralized Support Team, and a centralized Development Team.

Often, companies outsource application support duties first. No only does this make sense due to the changing dynamics driven by vertical and horizontal growth in IT, there may be considerable budget savings associated with outsourcing support. Also, many IT service providers have created very mature processes in the application support area allowing them to quickly transfer knowledge and provide value.

The reality of life in the large IT organization is that support work is often perceived as a non-glamorous role. I have often said of the Support Team "silence is their greatest praise." They are in the spotlight when something breaks but for the most part are ignored the rest of the time. As noted earlier, the traditional IT department typically follows a functional alignment with unique IT groups for procurement, sales, manufacturing, distribution, HR, finance, etc. Each functional group works independently. Together with their business partners they develop their own strategy and agenda for new projects; develop new applications and/or modify existing ones; and manage the support of existing applications.

Development clearly leads while support work takes a back seat in the traditional organization's hierarchy. This is a huge mistake especially for organizations with an ERP solution in place. Support should become an organizational focus with the same priority level as development and here is why: INCIDENTS DON'T JUST HAPPEN. The code doesn't stop working – electrons behave the same way every day. Every incident can be traced to a mechanical failure or some change that occurred somewhere. Perhaps it was a new operating system installed on a server; or maybe a minor modification was made to a line of code; or maybe someone entered an unrecognizable character string; or maybe file space is full. Remember, as the application footprint grows deeper with richer functionality, the horizontal flow of information will also grow. *The skill sets, processes and priorities required to manage the horizontal flow of information and find where breakdowns occur are totally different than those required for development.* Failure to recognize this will result in a poor user experience and rising costs.

Another reason to consider centralizing support is that support is rarely 100% of any one person's responsibility. It is often the case that nearly everyone in the traditional organization shares support work. The "curse" of being "on-call" rotates across the entire group with a different person identified for various time periods on a daily basis. On-call resources will be notified and expected to respond and resolve the issue when incidents occur – even during those off-hour times. Sometimes they get to sleep through the night, but more often than not, an incident occurs requiring their expertise.

Table 7-1 is a picture of what I call the Quality Spiral. It should not be a surprise that as complexity grows, so will the number of support calls.

Table 7-1 Quality Spiral

As more and more technology is implemented, complexity increases

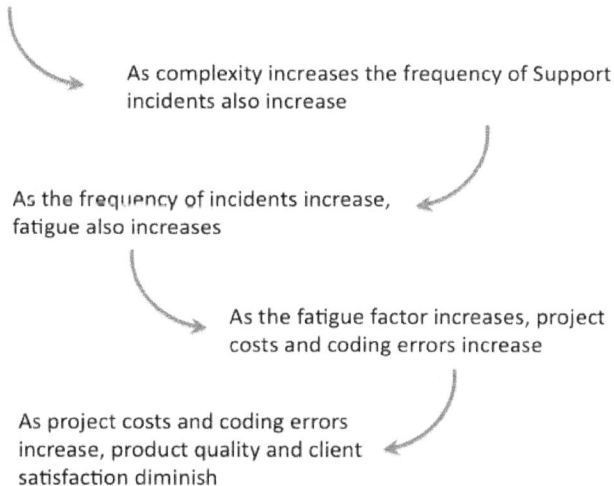

As complexity increases the frequency of Support incidents also increase

As the frequency of incidents increase, fatigue also increases

As the fatigue factor increases, project costs and coding errors increase

As project costs and coding errors increase, product quality and client satisfaction diminish

Unfortunately, the folks on-call during the evening also work on projects during the day. As the number of off-hour calls grows, there is an inordinate amount of fatigue and productivity loss. The person that responded to the incident also works on a project the next day!

Another issue is that because of time pressures and the need for sleep, the incident may be temporarily resolved and the root cause may not be corrected. A few weeks later, the same or a similar incident will occur again for a different individual. Up-to-date documentation is often lacking so there is no recorded knowledge of how it was solved before. This person will have to solve the incident from scratch without knowing if it is the result of bad code, a procedural error committed by the user or a disruption in the cross-functional data flow. Needless to say, in silo oriented IT groups Application Support doesn't get the focus needed to address faulty code, flawed processes and poorly trained users thus accelerating the "quality spiral" depicted in Table 7-1. Consolidating support resources into a centralized team will result in focused priorities, common processes, structure and behavior required to stop the quality spiral.

Outsourcing support in the traditional structure relied on a costly staff augmentation model with no vision or process. Outsourcing support using a Managed Services Model will drive fundamental process and cost improvements. I have found that moving all support resources into a centralized Support Team before outsourcing greatly improves results. Essentially, resources currently performing support will be identified and moved from their current functionally based team to a central Support Team. As noted earlier, few resources perform support 100% of the time in the traditional structure. This makes it difficult to easily determine who moves to the centralized organization. One needs to do some research to determine the number of hours devoted to support work and then determine the number of resources needed based on that. Exactly who those resources will be is a result of the negotiating skills of the centralized Support Team leadership.

The benefits of centralizing support are many and include:

1. Ability to design and implement common processes and standards across the organization.

2. Focused knowledge of how data flows and integrates horizontally between functionally oriented applications.

3. Ability to create a knowledge database containing information on application design, documentation, and incident resolution.

4. Key resources can be easily across multiple functional areas.

5. A central point to collect data and report supplier performance.

6. A single point of contact for the supplier, IT leadership, and business partners to interact with regarding Support concerns.

7. There is only one organization with which to clarify sticky points in the contract language and to modify or add language.

8. Centralizing the budget puts the cost of support out in the open so it can be easily measured and monitored. This often results in significant cost reductions.

How does centralizing support reduce costs? The support budget in the traditional IT organization is a nebulous animal. Resources were free to work on either support or development work. This is fine from a budgeting perspective as long as accounting rules are followed.

Here's what happened:

Accounting rules require development projects to be classified as capital expenditures since a tangible asset (the application) is created. Capital expenditures are amortized over five to seven years. Support work is expense and budget is approved year over year as a non-discretionary item. Managers would post resources budgeted as "support" as spending time on a development projects. Thus a portion of the support expense gets amortized and reduces the amount posted to the cost center budget creating a favorability. The issue is allowing a cost center to reporting both capital and expense.

Example: The budget for the sales systems function at XYZ Corp. is $3 million for 30 support resources plus $5 million for 50 development resources. The work activity assessment shows that only 10 FTE's or $1 million was actually needed for support. IT managers would post the time of the remaining 20 support resources to development projects thereby amortizing "expense". Amortizing $2 million of expense over five years is like an additional $10 million in capital bringing the capital total to $15 million – enough to fund 150 developers!

The original bottom line spending target for the XYZ cost center was $4 million (one fifth of the capital = $1 MM plus the $3 MM of support) funding 80 resources – 50 development and 30 support. Accounting games resulted in the ability to fund 160 resources (150 development and 10 support). The bottom line reported was still $4 million. One fifth of the capital is now $3 million plus $1 million for support. Sadly, the IT managers had no incentive to change this nor did finance since the budget objectives were met.

You can now understand why centralizing support is often met with vehement push-back. The spending and accounting described above is one of the primary reasons. The XYZ Sales IT group not only lost their $3 million support budget but because of the accounting games, they also went from an unauthorized resource base of 160 resources to 50. No wonder centralizing support generates so much interest.

Savvy CIO's see centralizing support as a catalyst to gain control of millions of dollars. The $2 million of budgeted support dollars that was used for development work is now exposed and can either be booked to savings or reallocated to strategic higher value projects. One of my clients had a support budget of $30MM and freed up $15MM after completing the transition to centralized support.

The first step in consolidating support into a centralized structure involves identifying the work that will be performed by the central Support Team. The second step is to identify staff and move budgets. Following our example, once consolidated, $1 million from the XYZ sales systems group will move to the centralized support budget along with 10 resources.

So how does one go about creating a centralized Support Team?

Timing and Objectives: Ideally, the transition should occur six to twelve months prior to outsourcing support work an MSM supplier. Create a small management team to lead the transformation. Identify a key executive to lead the transformation along with two or three leaders initially. The leader should report into a high enough level to be effective. Dedicated help from HR and Finance will also be needed to manage the transition process.

Once the transition team is formed, they will need to work with the incumbent resources to identify current support work elements and time required to perform them. The finance person can pull information on historical support budgets and headcount. Table 7-2 shows the support activities that should migrate to the centralized Support Team as the application moves along the path to maturity.

➢ Work elements for mature applications are easier to identify. Applications currently under development, recently rolled out, or experiencing upgrades, the total work transferred will be a function of where the application sits on its lifecycle curve.

➢ "New" applications have a core set of Support responsibilities that can migrate to the central team. However, the developers should manage the bulk of Support activities as noted.

➢ The "Invest" stage starts once the application is stable and known bugs have been addressed. The IT budget will fund enhancements over the next few years. The Support Team takes on additional duties noted in Chart 7-2.

➢ The "Mature" application is one that the IT team and Business Partners has decided to retire at some point and will no longer invest in. The Support Team basically manages all Support activities, including retirement, for mature applications. There may be changes for compliance purposes – either technical of governmental that the Support Team would perform.

Table 7-2 Application Lifecycle – Roles & Responsibilities

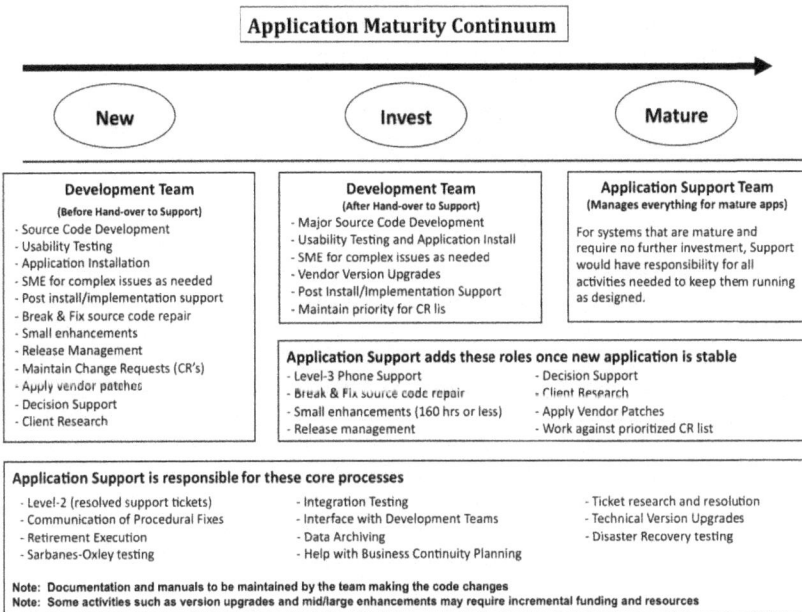

Application Maturity Continuum

New　　　　　**Invest**　　　　　**Mature**

Development Team (Before Hand-over to Support)	Development Team (After Hand-over to Support)	Application Support Team (Manages everything for mature apps)
- Source Code Development - Usability Testing - Application Installation - SME for complex issues as needed - Post install/implementation support - Break & Fix source code repair - Small enhancements - Release Management - Maintain Change Requests (CR's) - Apply vendor patches - Decision Support - Client Research	- Major Source Code Development - Usability Testing and Application Install - SME for complex issues as needed - Vendor Version Upgrades - Post Install/Implementation Support - Maintain priority for CR lis	For systems that are mature and require no further investment, Support would have responsibility for all activities needed to keep them running as designed.

Application Support adds these roles once new application is stable

- Level-3 Phone Support	- Decision Support
- Break & Fix source code repair	- Client Research
- Small enhancements (160 hrs or less)	- Apply Vendor Patches
- Release management	- Work against prioritized CR list

Application Support is responsible for these core processes

- Level-2 (resolved support tickets)	- Integration Testing	- Ticket research and resolution
- Communication of Procedural Fixes	- Interface with Development Teams	- Technical Version Upgrades
- Retirement Execution	- Data Archiving	- Disaster Recovery testing
- Sarbanes-Oxley testing	- Help with Business Continuity Planning	

Note: Documentation and manuals to be maintained by the team making the code changes
Note: Some activities such as version upgrades and mid/large enhancements may require incremental funding and resources

Reporting: The centralized Support Team should report into the position responsible for both support and development. The VP or Senior VP of Delivery (or similar title/position) is a good option.

Identify Staff: Once the work activities have been defined, specific individuals and/or FTE's can be named to transition to the central team. The transition may include internal and external resources. It is helpful to choose leaders for the Centralized Team that have experience in the IT functions to help sort the wheat from the chaff. The centralized team also needs some "stars" in order to function well – which can become an issue, as there never seems to be enough "stars" to go around. It is wise to keep the names confidential until HR and Legal have approved them and the organization is prepared to make the announcement.

Announcements: Be assured that most of the folks identified to move to support will not be happy. This is to be expected, as support has a number of negative connotations associated with it. However, with proper leadership after a few months they will not want to go back to the old model. Here a few helpful tips:

➢ Make sure the management team has been identified ahead of time and will be present for the announcement meeting.

➢ Minimize unknowns! Create a vision to share with the team during, or shortly after, making the announcement so they know what to expect. More importantly, they will know how they will fit in to the new organization and what they will be responsible for. When a person's job changes, they need to get through the shock and learn what their new role will be.

➢ Create a list of Frequently Asked Questions (FAQ's) that may be asked and the prepare answers to them ahead of time. A common question is: "When will *we* be outsourced?" One approach may be to slide by the issue by saying "there are no confirmed plans at this point to outsource any of your jobs." Of course, few will believe you and will think the opposite. Review your answers with legal and don't mislead the crew.

➢ Establish tactical plans and milestones the team can grasp and focus their efforts on. Schedule an all-team meeting to gather ideas to improve common processes and procedures. If possible, this could be done as breakout sessions at the meeting or take names of volunteers to serve on various teams and report back later. Themes should include: Improving team effectiveness, chronic recurring application issues that need attention, and how to work with a supplier half-a-world away.

  o Give the Business Partners a heads-up that the announcement is forthcoming and let them how their teams will be affected by it. Be sure to ask for their confidentiality.

  o Schedule the announcement and host a full-team meeting shortly after the announcement is made. Those affected will have lots of questions, addressing them quickly is important.

Potential pitfalls include:

➢ Trying to centralize and outsource Support simultaneously. It is best to centralize and stabilize the new organization first. Then optimize results by outsourcing the right work elements.

➢ Not getting the right level of staffing. The old functions are not always eager to reduce their head-counts.

➤ Not getting quality individuals. Make sure you receive some high quality folks along with the average employees.

➤ Transferees will be skeptical and generally negative at first, keep them motivated and recognize/reward positive actions.

➤ Business partners will likely be concerned with the change so it is wise to keep them informed.

➤ Assuring the funding for the staff being transferred is a key job of your finance person. Be sure to do a thorough scrub of cost center budgets for labor, travel, etc. as well as any software maintenance bills that move to the central Support Team.

➤ There is a common belief that Support can absorb incremental responsibilities without adding resources. Additional staff may be required to manage incremental activities since cost reduction efforts result in resource reductions for the original scope of activities.

One question I hear a lot is "should Support be recombined with Development at some point?" The answer to this question has to be NO. Any other answer will create a diversion and keep the new organization from solidifying and achieving their goals.

IT folks are usually not enthused about centralizing Support and see it as a negative change. In fact, many vehemently oppose it. The traditional IT managers don't want to give up any resources or power, and those at the lower levels dread a long-term career "sentence" of doing Support. The interesting phenomenon is that 6 months into the change, those in the central Support Team actually become enthusiastic champions of the change! Their attitudes improve as they create vital changes to the Support process and take pride in their ability to master the complexities of an integrated enterprise with data moving across and between application domains. People need to believe the change is permanent. Any discussions about recombining should be held in confidence at the CIO level.

Once centralized, the Support Team manages its own priorities. The results are impressive and include processes, cross-domain knowledge, and the time to resolve incidents all improve. Here is an example of how cross-domain knowledge minimized the cost and time required to make a change.

➤ I worked with a client that had to change a tax calculation for a European country they did business in. The estimate from Development was $200,000 requiring 4-6 months to complete. The reason for the high cost was that the change touched many applications across the enterprise resulting in complex testing requirements. However, the centralize Support Team managed the horizontal data flows daily and because of that, understood the integration layer very well. In addition they worked as a congruent team from end-to-end with a common goal. They took on the project and delivered the tax change after planning for one week and taking three days to implement and test the fix...ten days instead of 4-6 months!! The cost of the planning was $15,000; the cost of the code change and testing was $40,000; the response from the client was priceless!!

Once successfully transitioned, the centralized Support Team can begin to outsource their commodity activities to a Managed Services partner. The process improvements, user satisfaction and cost reductions associated with the MSM transition will be evident.

Some thoughts:

➤ The Support Teams staff will gain notoriety and (unfortunately) coveted for their knowledge and capabilities that resulted from setting their own priorities and improving processes.

➤ Recombining Support and Development quickly erode the benefits and process improvements gained by combining them in the first place. I have never seen those benefits continue after recombining, in fact, results get worse because Support loses its ability to set their priorities and become Development pawns.

➤ When support resources are re-absorbed by development, the improvements quickly fade away because staff are repurposed to work on development projects. Also, there is no focus on supplier management and the relationship with the vendor gets very ragged very fast.

Demand Management:

In today's IT world, it is often the case where development dollars and resources are not in sync with each other. There are instances where money is available but key resources are not; at the same time, we find resources sitting on the bench because money is tight. When individual functions create their own IT strategy and resource plans, they become self-focused. The essence of the Demand Management (DM) function is first to be the clearinghouse for all projects and IT spending and second, to balance funding and resources. The output of the DM team is the Demand Plan incorporating all project requests, costs, timing and resource needs from information into a central portfolio of IT projects. There are some very detailed planning tools in the market today. They show estimated delivery costs, cross-functional impacts and requirements for key resources, delivery timelines and expected benefits to the client. An IRR can be created for each project with an overall ranking assigned to it. Since the Demand Plan includes all approved projects across the enterprise, a quick roll-up of total capital spending, resources required and associated benefits can be aggregated for upper management.

Watch outs:

> Prioritizing how to spend the IT budget is the goal. This can be done based on IRR calculations within the planning tool or other metrics. However, clients with the loudest squeak will be end up being well greased. Business partners have their agenda and fight for their projects regardless of what the IRR says. After all, the way the accounting works in most IT shops is that the ongoing cost of a new application (support, maintenance and depreciation) does not directly hit the client's books. The client benefits from cost reductions that are essentially free. (It is no wonder they have a robust project agenda). These costs are absorbed in the IT department along with the task of explaining why IT costs continue to rise every year at a 10-15% clip. Some leading edge firms are working to directly charge the client budgets for the ongoing costs. After all, the benefits received from the project should cover incremental expenses. It is a good way to create a more accurate benefit number.

> Data collection can be burdensome. The more complex the planning tool, the more scrutiny it requires from the IT project managers. This can result in shortcuts that threaten the integrity of the planning tool itself.

> Concurrent resource planning across multiple functions is difficult to achieve. There are certain skill positions needed for every project. An example would be Database Analysts. DBA's are in short supply so it is important to know how much of their time will be needed for each project and when that need will occur. It is not unusual to have 100 projects, or more, in the plan at any given time. Trying to align the DBA resources against an annual project plan is daunting. When the Demand Plan changes (and it will change frequently) the resource plans needs to be reset as well. In addition to the DBA role, there are many others that require cross-project coordination as well.

Demand Management (DM) doesn't really require a structural change to the IT organization. What's needed is an IT-savvy person with some financial smarts and the ability to work within a structured database.

Senior members of the CIO staff will own the DM process acting as a "board of skeptics" routinely reviewing the plan to make sure the project leaders have thought through all aspects of the project and raised the right questions.

Flexibility and responsiveness are needed to ensure compliance to the plan. No projects should start if they are not on the demand plan. There are exceptions such as structural repairs needed for incident resolution, or testing when a technology stack is upgraded – the Support Team can manage most of these. One way to enforce this is to expense all projects that are not on the approved plan. As is the case with most plans, once set, something will happen that requires it to change so it is important to be able to add and/or subtract projects from the plan quickly to maintain it's integrity.

Help Desk and Call Management Centers:

Help Desk and Call Management Centers are key enablers of successful outsourcing. There are a number of firms that specialize in providing the technology and call center agents on a fee basis.

The Help Desk (HD) exists as a physical place for users to call when they experience a service failure with an application or to request a new service (adding an application to their workstation). Service failures (incidents) must be managed and resolved quickly in order to restore service. The Help Desk is staffed with individuals trained to follow a script. The script is designed to ask a series of questions to help better identify specifics about the incident thereby making it possible to solve the problem while on the call or route it to the Support person's queue for resolution.

Call Management Center (CMC) is the processes and/or tools used to manage the calls that come into the Help Desk. The two terms can be interchanged and often are. When the user calls, the HD agent gathers identifying attributes from the caller (name, phone, location, etc) and asks a series of scripted questions. Based on the response, the HD agent branches through the script to a resolution or will open a ticket and route the call to someone with a deeper understanding of how to resolve the incident. The HD agent can automatically route the ticket to the queue of the Support Team responsible for resolution. The queue is an electronic mailbox accessible by the Support Team. The ticket is opened and the Support person reviews the description of the incident. The Support person may need to contact the user directly for more information and/or can walk the user through the steps necessary to fix the problem. When the incident is resolved, the Support person closes the ticket by identifying the cause of the service failure and what was done to remediate the incident.

The ticket associated with the incident remains in the CMC system database forming the foundation of data and metrics to improve service quality and manage the supplier. Here are some examples:

➢ *The date and time the ticket was opened.* This is the starting point of the incident and time-related metrics refer to this time stamp.

➢ *Applications and/or processes associated with the incident.* This helps identify applications or procedures that are chronic problems helping to establish where to focus improvement.

➢ *A brief description of the incident.* This is needed to give the HD agent, through scripting, the best chance of routing the ticket to the correct queue the first time.

➢ *The queue the ticket was routed to.* This is vital to being able to understand relative staffing levels needed on the Support Team. A queue is associated with an application group. If the queue for the Financial Systems gets 100 tickets per day, they probably need more Support resources than an HR Systems group that only gets 20 tickets per day in their queue.

➢ *The time the ticket is first opened by Support.* Response Time is a common SLA when outsourcing Support. The time stamp created when the Support staff acknowledges the ticket vs the time stamp when the ticket was opened is the response time. If the SLA is 15 minutes, the Support Team needs to be staffed to make sure every ticket is acknowledged in 15 minutes or less.

➢ *Captures "misroute" errors when the ticket goes to the wrong queue.* Misrouted tickets generally result from incomplete or erroneous scripting. At any rate, when misroutes increase, one needs to review scripting to assure that the HD agent has the best information available.

➢ *Identifies queues when other groups are needed to help with service recovery.* This is important as it tracks the root cause of incidents that occur outside of the application. For example: the user call indicates an issue with the Order Management system and so the ticket is routed to the Order Management queue. The Support person does initial research on the ticket and realizes that a database is experiencing contention causing the application to hang. The Support person contacts the DBA to help resolve whatever is causing the contention. The ticket will go to the DBA's queue assuming they have one.

➢ *The time the ticket is closed.* This is an important metric. One of the basic SLA's is how long it takes for the supplier to resolve an incident and restore service. If the SLA for resolution is 4 hrs, the time stamp when the ticket is closed is compared to the time stamp when the ticket was opened. The difference is the resolution time.

➢ *Description of how the incident was resolved.* This is useful when the incident repeats itself in the future. The HD agent and/or Support staff can quickly search the database looking for similar incidents and how they were resolved in the past.

➢ *Description of the root cause that resulted in the incident occurring.* Service recovery and fixing the root cause may be two separate events. The immediate concern is to restore service so the user can do their job. Sometimes, a permanent fix is applied at the same time service is restored. However, some root causes may take more time to plan, code and install the application fix to permanently resolve the incident. The analogy is to place a bucket under a leaking pipe to control water damage until the plumber can repair the leak.

➢ *The time it takes the root cause to be corrected.* Once a root cause is identified, the clock starts ticking on how long it takes to correct it. This becomes an important metric for another SLA for the supplier to achieve.

Many CMC vendors provide software supporting an automated self-service response for users. A good example is resetting passwords on line. This results in a more rapid response for the user and reduces the workload for the HD agent, which in turn, reduces cost.

The other role of the HD is to collect Change Requests (CR) from the users. A user may need help installing a new printer. As with incidents, a ticket is opened and goes to the queue of whoever is responsible for installing printers. CR's can be for hardware, software, or functionality additions to existing software.

The organizational changes discussed in this chapter will enable a successful outsourcing experience. Detailed plans will be needed to keep the project on its path and avoid costly delays. Project Objectives and Milestones will be discussed in the next chapter.

Chapter 8:

Project Planning

Table 8-1 provides a list of common strategic concerns along with examples of various tactics and actions when creating the project plan. We will expand on Project Definition and Knowledge Management in this chapter. People impact and SRM will be addressed in Chapters 9 and 10. Implementation will be covered in Section 4 of this book.

Table 8-1 Project Objectives and Milestones

Strategic Concerns	Tactical Focus
Project Definition	- Vision/scope and major milestones - Big bang or waves - Impact to IT landscape and organizational alignment - Management updates
Knowledge Management	- Capture, store and retain existing and new knowledge - Subject Matter Experts - SME - Sharing knowledge with other parts of the organization
People Impact	- Identify impacted employees and separation dates - Determine severance costs - Look for redeployment opportunities - Collateral damage
Supplier Relationship Management	- Request for Proposal (RFP) - Master Services Agreement (MSA) - Key Metrics (SLA, KPI, etc) - Sourcing model - Payment terms
Implementation	- Detailed project plan - Knowledge Transfer (KT) - Preparation for "go live" - Continuous improvement - Governance reviews

Project Definition:

Once defined, there should be no doubt in anyone's mind what the vision and scope of the project are. An example of a vision statement would be "Outsource all application Support activities to a third party IT service provider achieving a managed services solution." The vision must clearly define what is to be achieved as a result of the project. The scope will provide boundaries and help focus the team's efforts in the right direction. The scoping document is important and should be maintained over the life of the project to capture changes to it. If outsourcing Support is our vision, the scope will define which IT functions are included along with specific applications and Support activities the supplier will perform. For clarity, it should also identify the applications and activities that will NOT be included. Make sure there is verbiage about who will be managing the out-of-scope items and how the supplier will interact with them. Being clear about the boundaries will eliminate confusion down the road.

If one of the project objectives is to generate cost reductions, it is important to understand how much savings will be achieved and by what point in time. In many cases, the savings stream from the cash flow analysis provides a foundation for the "how much" and the "when." However, a tactical plan is required to define the specific actions required to assure the savings are achieved. We need to get down to the details of which work activities will be outsourced (and which will not). It is time to create a confidential list of impacted employees by name; identify technical upgrades that may be needed; define supplier roles & responsibilities; determine best geographic location(s) for offshore resources; understand how the supplier will interact with other IT providers; consider how the project will impact the broader IT organization; align project milestones with business cycle timing (i.e. you don't want your project to go live during quarter end); and many more.

Milestone Plans:

The overarching strategic plan for the project will identify major milestones and goals. Clearly defined goals, and dates for achieving them, will help the project team stay focused as they dispatch the initiative. Project milestones should be aggressive yet achievable. There are many steps associated with implementing a project and delays at any point may well impact delivery of the overall plan. When delays become imminent, they must be raised quickly and solutions found quickly in order to keep the project on plan. Examples of major project milestones and sub-tasks include:

➢ Finalize Working Team (WT) members by _____ (date).

- o The Executive Steering Team (EST) identifies the IT core group to work through all phases of the project becoming the nucleus of the governance team. The EST needs to make sure they are free of existing responsibilities so they can focus 100% of their time on the outsourcing project.

- o Identify representation from Finance, HR, Legal and other functional areas as appropriate for the project.

- o Identify temporary IT and/or client Subject Matter Experts (SME). The SME's will provide the knowledge that will be transferred to the supplier resources. They will be scheduled "as needed" through the implementation phase.

➤ Obtain EST approval to proceed by _____ (date).

- o Clarify the general vision and scope of the project. This is critical and must be done before the cash flow analysis.

- o Complete preliminary activity/work analysis to identify the activities and applications that will be in-scope. Determine the person hours and FTE's required to complete these tasks.

- o Create a timeline for realization of savings showing anticipated spending increases as well over the course of the project. Complete the preliminary cash-flow analysis.

- o Decide on what to measure to determine if the promised savings have actually materialized. This can be problematic as the project may likely be a subset a larger cost center. Assigning the project to it is own cost center will make "before" and "after" comparisons easier to make.

- o Prepare/present the review document to the Executive Team for approval.

➤ Launch the RFP and select supplier by _____(date).

- o Create a well-defined scoping document that will become the backbone of the Request for Proposal.

- o Arrange applications into logical groups such as: Financial; HR; Supply Chain Planning; Manufacturing; Procurement; Supply Chain Execution; etc. to facilitate supplier review.

- o Anticipate what the supplier will want to know to accurately bid on the project. Examples include: technologies used; number of users; incident details (number, frequency, duration), current staff supporting those applications in scope.

➤ Identify Process Team (PT) resources by _____(date).

- o The PT is client staff and supplier onshore and offshore resources. The PT is vital in on-going governance as well.

- o The PT will be responsible for training to resolve knowledge gaps; monitoring supplier performance vs SLA's; escalating incidents that exceed time-to-resolve criteria; acting as liaison between supplier and client and/or end-users.

- o The WT determines start and end dates for implementation. Planning 20 weeks for this is aggressive but doable with strong leadership and aligned priorities.

- o Project Planning: allow 4 weeks to assign project roles; identify Subject Matter Experts (SME); schedule KT sessions with SME's and new resources; offshore resources travel to U.S.; VISA processing; establish connectivity with offshore resources; etc.

- o Knowledge Transition: allow 4-6 weeks during which SME's work with new resources to train and transfer existing knowledge to them.

- o Secondary Support: allow 4-6 weeks for the new resources shadow existing resources to document all daily activities.

- o Primary Support: allow 3-4 weeks during which the new resource actually performs the required activities with the incumbent backstopping them.

- o Steady State: occurs at end of Primary Support phase with new resources taking over on their own. Incumbent resources can now be redeployed.

Big Bang or Waves:

When developing strategic plans, one needs to decide if the implementation will all happen at the same time (Big Bang) or be comprised of multiple threads that will each be implemented to their own schedule (Waves). The primary factor for deciding which to do is the size of the effort compared to the number of resources (staff and third party) available to work on it. In most cases where outsourcing only touches a small handful of applications, doing them all at once may be more efficient. The Working Team should make the decision, as they will be much closer to the issues, the work to be done, and available resources. I've done both and, with proper preparations, success can be achieved. Waves are nothing more than a series of "big bangs" sequenced in a logical manner.

Table 8-2 is an example of the transition plan for a very large project. The team decided to break the project into 4 waves in order to minimize over-loading key resources. Within each wave, major application clusters were identified with a transition plan created for each one. Depending on the complexity of the cluster, there is generally 6-8 weeks dedicated to Knowledge Transfer, 4-6 weeks for Secondary Support and 4-6 weeks for Primary Support. Although not shown on this chart, each week within each cluster is broken down to a daily plan showing scheduled KT sessions for SME's and supplier resources. The supplier will definitely have a point of view on this but remember, the client SME's know the applications best and can offer unique perspectives especially for homegrown applications. We will go into greater depth on the implementation processes in a later chapter but creating and tracking the daily transition activities is something top tier suppliers do very well.

Table 8-2 Implementation Plan Overview

Wave	Sub Waves	Application Cluster	Week 1	Week 2	Week 3	Week 4	Week 5	Week 6	Week 7	Week 8	Week 9	Week 10	Week 11	Week 12	Week 13	Week 14	Week 15	Week 16	Week 17	Week 18
1	1A	Procurement	K	K	K	K	K	K	K	S	S	S	S	P	P	P	P	C	C	C
	1A	Transportation	K	K	K	K	K	K	K	S	S	S	S	P	P	P	P	C	C	C
	1A	Administration	K	K	K	K	K	K	S	S	S	S	S	P	P	P	P	C	C	C
	1C	Time Clocks			K	K	K	K	S	S	S	S	S	P	P	P	P	C	C	C
2	2A	Allocation & Replenishment	K	K	K	K	K	K	K	S	S	S	S	S	P	P	P	P	C	C
	2A	Inventory	K	K	K	K	K	K	K	S	S	S	S	S	P	P	P	P	C	C
	2B	Warehouse Management		K	K	K	K	K	K	S	S	S	S	S	P	P	P	P	C	C
3	3A	Product Design	K	K	K	K	K	K	K	S	S	S	S	S	P	P	P	P	P	C
	3A	Financial Planning & Analysis	K	K	K	K	K	K	K	S	S	S	S	S	P	P	P	P	P	C
	3B	Revenue Forecasting		K	K	K	K	K	K	S	S	S	S	S	P	P	P	P	P	C
	3C	Cost Accounting			K	K	K	K	K	K	S	S	S	S	S	P	P	P	P	C
4	4A	Field Operations	K	K	K	K	K	K	K	S	S	S	S	S	P	P	P	P	P	P
	4A	HR Benefits, Payroll, etc	K	K	K	K	K	K	K	K	S	S	S	S	S	S	P	P	P	P

	Key	
K	Knowledge Transfer	
S	Secondary Support	
P	Primary Support	
C	Complete - Steady State	

Competing Initiatives:

Nothing happens in a vacuum, especially in IT organizations. It is wise to take the time to think through how the outsourcing project will impact the IT landscape. Identify other major initiatives that will be competing for the same resources that are needed for the outsourcing initiative. If major projects are in play, perhaps they can utilize some of the internal staff that outsourcing will free up.

Let's assume that in addition to your outsourcing project, you find there is a major implementation of SAP planned by different part of the organization. Sally is working on the outsourcing project for you but will be let go at the end of it. She has the skills and experience needed that make her a good fit for the SAP assignment. Moving Sally to the SAP team would be a great solution as it provides expertise needed by that project and eliminates a layoff. It also keeps Sally productive and positive while working to complete the tasks for the outsourcing project. Unfortunately, getting all the tumblers to align on when Sally can move is difficult. For example, the project may need to keep her through the end of knowledge transfer, still 8 weeks away. Unfortunately, the SAP team has a need for her next week. Being careful not to shortcut the KT process, you really can't afford to release her early as it will cost you down the road. At the other end of the spectrum, there may be a project that won't need Sally's services for 8 weeks after she is done with yours. Holding on to her for those additional weeks will impact the project cost. I have led projects that successfully transferred hundreds of impacted staff to other work in order to avoid layoffs. The effort involved is worthwhile.

Here are some ideas to consider:

➢ Start this planning process as early as possible. Don't wait until a week or two before Sally is available.

➢ Identifying someone else on your team or the SAP team to "fill in" for Sally to adjust the timing of when she can be released.

➢ Work with the SAP team to determine if Sally's time can be shared across both projects.

➢ Work with your supplier to see if they have anyone that could bring some of the skills Sally has to either project. Of course you will need to work through the details of which project pays for the external help.

➢ If you need to hold Sally longer than planned to fit the timing of the other project, seek to get a waiver of sorts from the finance team. It may be far less costly to pay her for those additional 8 weeks than to lay her off. It is important to keep a close watch on costs to avoid surprises at the end of the project.

➢ Investigate options to adjust the wave timing to avoid conflicts.

Project Updates:

Once the project moves into implementation mode, there will definitely be a need for frequent project updates. The many unforeseen variables that always seem to pop up can cause delay. Often times the EST and WT are able to help resolve issues but they need to be kept abreast of concerns and not be surprised by them.

The EST should create a format for what gets reviewed with them, the frequency they want to be posted, and who should attend. The WT would be requesting the same of the Process Teams. Here are some thoughts on content, structure and frequency:

➢ The Executive Steering Team should meet monthly with the WT to assure that the project moves through the exploration, approval and implementation phases as quickly as possible.

➢ Once the project has achieved "steady state" (meaning it is complete) the overview with the Executive Steering Team can change to quarterly pace. Topics to cover include:

- A review of relationship quality.

- Opportunities and suggestions for process improvement.

- Supplier suggestions for innovation and productivity.

- Miscellaneous concerns from the project leadership.

- Typical membership would include the EST, the WT lead managers and the supplier's relationship manager.

➢ The Working Team should meet weekly (or as needed) during the project planning and implementation phase to address all strategic issues that arise during outsourcing.

- A monthly pace is fine after Steady State is achieved to review SLA's; status on Minor Enhancements; technical concerns; security; progress on initiatives; etc.

- Typical membership would include the WT, Process Team leads, and supplier leaders. Clients should also be included.

➢ Process Team members will be working in a closely collaborative atmosphere with each other. The means few "formal" meetings are required.

- Once steady state is achieved, Process Teams should meet weekly to review incident patterns, monitor progress on minor enhancements, address training needs, resolve personnel issues, etc.

- The Process Team consists of all supplier and retained client staff. These are the "hands-on" folks that will perform the tasks associated with the project.

Knowledge Management:

Knowledge retention, or more precisely the loss of it, is by far the single greatest risk associated with outsourcing. Knowledge Management is the collection of processes that govern the creation, dissemination, and utilization of knowledge. Firms need to take action to prevent the loss of internal knowledge about the detailed technical nuances of their environment. In today's world, most users rely heavily on their IT capabilities for the completion of even the simplest daily tasks. Once the technical knowledge about an application is lost, the cost of getting it back will be sizable; assuming it can be recovered at all. It is somewhat easy to find resources that understand technologies – communication protocols, coding, servers, etc. – but how those technologies link to the applications and business rules are essentially unique for every company. Any savings achieved through outsourcing may soon be lost if the knowledge is lost. Someone needs to take responsibility for creating the Knowledge Management solution for your firm.

Outsourcing IT services will accelerate the firm's knowledge erosion as the supplier becomes more deeply imbedded within the organization. As the size of the supplier's footprint grows, the small contingent of retained staff will need to focus more and more of their time on managing the relationship to assure that SLA's are met and the value proposition is delivered. Things will be fine for a while. The supplier will quickly learn each application and develop the expertise needed to resolve any incident that occurs. This is good for keeping the lights on and maintaining status quo. At some point, however, the application may need significant modifications. If knowledge has not been proactively maintained, there may not be anyone left that knows enough about the application to manage modifications. Unfortunately, the alarm bells won't go off until it is too late. Risks associated with knowledge loss include:

➤ Inability to know if a supplier's work estimate is reasonable.

  o Knowledge will erode to a point where the IT organization can no longer determine if a work estimate is reasonable. (ie. does the estimate include all touch-points or not?)

  o Predicting unintended consequences will be compromised. If application "x" changes, what changes with application "y"? These changes will impact testing costs.

➤ It will be difficult to reduce Total Cost of Ownership (TCO) because knowledge of the environment is lost and knowing where to drill for cost reduction opportunities will be compromised.

➤ Increased risk of being held hostage by the outsourcer. Once the knowledge is lost internally, how will you ever transition to a new supplier?

Fortunately, there are a number of tools and techniques available to help establish a Knowledge Management solution. Many of these tools are computer based and boost the effectiveness of managing explicit knowledge by making it available to those who need it. Here are a few examples:

➤ Start by identifying what information to capture, how to organize and store it, how to define it (metadata) and who the gatekeeper will be.

  o Examples of what to capture: application documentation, code schema, context diagrams, environment changes, application or infrastructure upgrades, incident reports and resolution, etc.

➤ Consider implementing groupware technologies that will be able to facilitate web-conferencing and sharing; along with internal intranets to facilitate database searches. These tools contain search engines, alerting, intelligent agents and data mining.

Project success is directly correlated to the quality of the planning. There are many issues that will come up during the course of the project each trying to push the effort off course. Don't let that happen. I managed a complex global transformation for a large ($10+ Billion) client. We spent 3-4 weeks creating the detailed work plan and executed it flawlessly even in the face of many internal and external obstacles. The team was motivated to figure out how to keep these roadblocks from derailing the project. To the team's credit, we implemented a major transformation in only 5 months – from contract signing to steady state. Solid planning on the part of the supplier and my staff were the foundation for this success.

Unfortunately, one of the unpleasant results of outsourcing will be the elimination of incumbent staff. How this is done speaks volumes to the character of the organization. The next chapter is dedicated to the disposition of impacted staff.

The *Inside* of *Outsourcing*

Chapter 9:

Disposition of Impacted Staff

Lets deal with the elephant in the room: Outsourcing will result in job eliminations. In order for outsourcing to reduce cost, existing people need to come off the organization chart. There is no pleasant way to deal with this aspect of outsourcing. However, it should be done in a thoughtful, professional manner. I can't emphasize enough how much an experienced HR partner will facilitate the activities discussed in this chapter. Table 9-1 outlines the strategic concerns associated with managing impacted staff.

Table 9-1 Impacted Staff Analysis

Strategic Concerns	Tactical Focus
Define Workgroups	- Work going to outsourcer - Jobs being eliminated and who is performing them
Determine Magnitude	- Identify how many employees will be affected - Review employment law and impact on local economy - Severance costs
Identify Names	- Who is performing the jobs to be outsourced - Create job and skills inventory - Use blind analysis method to keep names confidential
HR and Legal Reviews	- Provide rationale to defend who was selected - Make sure the selection process is consistent
Financial Preparation	- Severance cost and budget - Where will budget impact fall - Prepare severance packages
Placement Options	- Redeploy to other projects - Rebadging affected employees
Notification and Timing	- Confidentiality is key as dynamics change frequently - Timing of staff departure and when to announce it
Organizational Reaction	- Consider the reaction of those that are not impacted - Try not to drag things out

Define the Work Group:

In prior chapters we discussed the need to identify specific work activities to be outsourced. This exercise resulted in a reasonable estimate for the number of incumbent resources that may be impacted by the outsourcing project (an "impacted" resource is someone whose job is being eliminated). Some of those impacted may be contractors, while others will be employees. Some might be redeployed to different jobs within the organization while others may end up leaving. I have managed a number of transformation projects in which people lost their jobs. It is an unfortunate outcome of outsourcing. It is neither easy nor pleasant to select the list of impacted names. These are people you have worked with over the years and are more than just "names." Creating the list of those impacted is more complicated than just connecting people to the tasks that will be outsourced – we also need to plan for and minimize the amount of "tribal" knowledge that will be lost when they walk out the door.

I find it helpful to start by defining organizational boundaries for work activities and employees that may appear on the outsourcing "radar screen." The workgroup that will be impacted is found within the boundaries of this defined universe. For example, if it were decided that the project includes all application support work for financial systems, the work group would include those currently on the Finance Systems Support Team. As we discussed in Chapter 7, this is much easier done when support has been consolidated into a centralized team first. Otherwise, the entire Finance IT team including both development and support resources forms the universe. This is certainly the case where support duties are rotated across the group. I have managed major projects designed to outsource all application support and in each case, support resources were first consolidated into a centralized team and organizationally separated from their old teams. Had we not done that, the entire IT organization becomes the universe by default. The project would have taken significantly longer in order to vet through the roles of each individual across the entire department. Prepare to explain the strategy for how workgroups were defined and review it HR and legal.

Once the work activities and workgroups are defined, the next step is to determine how the work activities relate to existing jobs. A job is the cumulative responsibilities for activities performed by one or more individuals. The jobs that closely align with the work that will be outsourced are identified. I was fortunate to have worked with great HR partners who spent a great deal of time rewriting job descriptions to reflect the contemporary work environment. This is an important distinction: it is the individual jobs that will be outsourced; the people doing those jobs will be redeployed or let go. In cases where multiple people are performing the same job, seniority is used to determine who stays when a portion of the work is retained; assuming all other factors are the same.

The focus of this exercise must be on the jobs being outsourced, and not on whom we would like to get rid of. Employees with a documented performance plan should be dealt with independent of the outsourcing project. The operative word is "documented." Avoid the temptation for outsourcing to be a means to eliminate poor performers. Here is an illustrative scenario that happens too often.

> ➢ Although this story is illustrative, I am sure you can relate to it: One of the IT team managers, Fred, was outsourcing a large percentage of his team. He made sure that Jill was eliminated as he had a difficult time managing her. Everyone knew her work was below par and her personality was difficult to work with. She did, however, have more seniority than many of the others doing similar jobs. Fred decided to use the outsourcing project as a reason to clean house and get rid of her. Unfortunately Jill had an empty file. Fred not only failed to document Jill's performance, he had never sat down and discussed her personality or performance issues with her. HR and legal conducted a staffing review with each manager to assure their cumulative actions were consistent and would stand the test of time. Needless to say, without any documented corrective action plans, Fred was not allowed to eliminate Jill. He was forced instead to eliminate Bobby. Bobby was in the same workgroup as Jill. He was fresh out of college and had a lot on the ball. He was easy to work with and picked things up quickly. He never complained about working extra hours (without pay) to get the job done. Oh yeah, since he was a recent college grad he therefore had the lowest seniority in the department. He is still missed today. Fred (and all the other Fred's out there) failed miserably as a manager by not conducting performance reviews with Jill. He was not doing his job. Managers should have conversations with people that are not meeting expectations. These discussions should be documented showing the progressive levels of disciplinary action taken. If Fred had done this, Jill would be gone and Bobby would still be on the payroll. Make sure bad apples are dealt with properly.

If there was one thing I learned from my HR partners it is this: use a consistent process and be consistently fair when selecting the impacted individuals. With the emotional impact job eliminations have, it is not unusual for someone to seek litigation after the fact.

Determine the Magnitude:

Once the work group has been defined, it is typical that about 70% - 80% of the existing team will be impacted. Those retained will either be redeployed to other IT jobs or form the nucleus of the governance team. Sometimes the work group only includes employees but often it is a mix of internal and external resources. One of my clients outsourced about 200 resources in their centralized support group. One hundred and twenty five were employees and the others were third party contractors. The magnitude of the entire effort is important, but what matters most is the number of internal employees affected. Once the impact has been estimated, work with HR to determine the extent to which local employment law will affect the timing and cost of your project. For example, some states require at least 60-days notice if the workforce reduction is greater than 50 employees. Your HR partner will be able to provide this information for you.

Think about the impact a layoff will have on the local economy. A reduction of 100 people in a small town will likely create a larger stir than a layoff of 500 in the Chicago area. Be prepared for how your firm will respond to the questions from local media. Your PR staff (or whoever is in charge of this) should prepare talking points for the local media. And don't forget the impact within the firm too. It is wise to create a list of Frequently Asked Questions (FAQ's) and talking points to share.

Identify Names:

Once the jobs that will be eliminated have been identified, the next step is to match names to the jobs being eliminated. Although it sounds simple, matching names to jobs can be complicated by the fact that many jobs are loosely defined, and may have elements that are not captured in the existing job description. Job descriptions should be maintained frequently but unfortunately go to the back burner when HR staffing is cut. However, it is important that they be made as current as possible. Complicating matters further, some employees perform certain tasks that are not included in their primary job's description. These ancillary activities need to be factored into the decision making process as well.

Another preparatory step is to assure that the "skills inventory" for each employee is up-to-date, especially for those on the list of impacted people. Skills inventories require a great deal of effort keep current, which is typically its Achilles heel. The skills inventory will show the depth of knowledge and experience of those impacted. It also becomes a good information source for other managers that are looking to fill open positions on their teams. Think of skills inventory as an employee's "mini" resume.

They should reside on an interactive database so each person can easily update their own information. Using a searchable database will make searching and sorting easier. On a routine basis, employees should be prompted to make updates to their personal information and experiences.

One approach to help assure a non-biased selection of impacted staff is to create what I call a "blind analysis." For employees in the defined workgroup simply replace their name on the skills inventory with a unique identifying number. Ask your team to evaluate which "numbers" most closely fit the specific jobs being eliminated. Not only does this offer a more unbiased solution, it also helps to keep the names of those impacted confidential.

I can't stress enough the need to keep the list of names confidential. The main reason is that the jobs being eliminated and the individuals performing them may change during the evaluation period. People change jobs frequently, new work may be assigned to existing jobs, and the scope of the project may change. Once the selection process is finalized, schedule a confidential review with HR and Legal to assure that there are no potential issues with the list. Waiting until the last minute to announce the names of the impacted staff is a good thing, in fact, there is nothing to be gained from a premature leak.

HR and Legal Review:

This review will offer the manager an opportunity to defend those selected for layoff. Be prepared to explain how the workgroup was defined and how the specific jobs being outsourced relate to the workgroup. Also, refer to the skills inventory, and any other associated methodology as evidence that an unbiased consistent process was used to create the final list. The review session may raise issues with regard to any bias evident in the final selection such as age, race, etc. It might be difficult to defend a list that is primarily employees over the age of 50, or women or persons of color, or any other protected class. This review is an important exercise to eliminate issues down the road.

Financial Preparations:

Severance is the salary and benefits paid to each impacted employee when they are laid off. The size of the severance package can range from nothing, to a few weeks of pay, to a full year's salary or more and is totally based on internal HR policy. The actual severance paid may be different for each employee as it is typically based on years of service and salary grade within the organization - VP's generally have a more robust severance package than those in a lesser pay grade. Severance costs may be the single largest incremental expense associated with outsourcing. Because of that, it is important to generate an estimate for the total cost of severance. HR and Finance can provide data to help with this estimate by looking at the average costs and years of service within the workgroup. This is an important number for the cash flow analysis. It is fine to run the cash flow model using anticipated severance costs to understand how it affects the model. Run the model again once actual severance costs are known.

The more important question to answer is "which budget will cover the cost of severance?" Some companies will create a financial reserve to cover the cost of restructuring, which generally includes severance. Establishing a reserve is quite complicated and must meet accounting guidelines. It will take time to create the reserve so don't wait for the "last minute." Clearly, one of the roles of the executive team is to secure funding to cover these expenses.

In lieu of a reserve, it is possible to "self-fund" the cost of severance and other one time costs with project savings. If projected savings are $200,000 per month, and incremental cost (including severance) is $1.6 million, it is possible to offset (or self-fund) the expense with 8 months of project savings. This of course means that the project must be completed during the 1st Quarter of the fiscal year. This is do-able but difficult and requires solid about project timelines. One option is to start the project in the 4th Quarter of the prior year with completion targeted for some time in the 1st Quarter. Severance payments for terminated employees don't start until after the employee leaves the company so in this example, severance hits the books in the 1st Quarter making it possible for the projected savings to offset the cost. Work with your finance partner to develop a reasonable approach to minimize negative budget affects.

Once the names are finalized, Finance and HR can work together to create a unique severance package for each impacted individual. This document describes the termination date, severance pay, and benefits. The document may also describe actions the employee must take in order to receive the severance payout. For example, the termination date is considered to be the point in time when severance begins. The expectation may be that the employee will continue to perform his/her job until the termination date in order to minimize disruption. This gets sticky in some cases because:

➤ The terminated employee may find a job elsewhere before they have trained their replacement.

➤ If the employee does not perform their job, as they may be denied severance and terminated immediately for cause.

➤ Lucrative severance packages are an incentive for the employee to stay and continue to perform their job.

- ➤ Where the severance amount is small, a "stay-pay" bonus may be used to entice the employee to stay and perform their job through transition. If they perform adequately, the bonus is paid to them on termination. When stay-pay bonuses are used, they should be well defined and part of cost of severance.

- ➤ In some cases, the organization may ask the employee to stop coming to work immediately but continue to pay them until their termination date is reached. This minimizes any ill affects caused by keeping the severed person in the office.

Placement Options:

Outsourcing may be viewed more positively when impacted employees actually move to other work within the firm. The project will benefit from reduced severance exposure and morale will be better. Unique talent remains in-house so they can be tapped later if needed. Here are some ways HR may be able to help with this.

- ➤ Instead of paying for severance, agree to fund these employees until they are placed within the organization. This may erode project savings but is worthy of consideration if the number of impacted employees is manageable.

- ➤ Give impacted staff the first option to post for open positions.

- ➤ Work with various project leaders to see if the affected employees have skills that could add value to their projects. Remember, these folks are being laid off because their job was eliminated, not because of poor performance. Many have skills that are useful in other parts of the organization.

- ➤ Some will be good candidates for roles on the Governance Team that will manage the relationship with the supplier – especially those with a broad understanding of how the applications and business processes work together at the integration layer.

- ➤ Rebadging is another option where the outsourcer agrees to hire impacted employees. It is called "rebadging" due to the fact that the affected individual will become an employee the outsourcing company and wear their ID badge.

  - o Rebadging may reduce severance cost depending on how your HR department has structured the rules about severance. Since no job loss actually takes place, severance is usually not paid.

    - ▪ Rebadging can be beneficial to the client if the rebadged employee does the same work they did before often in the same physical building. In this way, the client doesn't risk the loss of job knowledge.

- The savings from rebadging is based on the blended cost of resources working offshore and those that are rebadged staying onshore. Understanding the cost impact of rebadging is critical.

o India-based outsourcing firms will not generally advocate rebadging. They know the savings, and their profits, come from labor arbitrage driven by pay rates in India. They also adhere to well-developed internal processes for managing the work. They know that hiring the client staff will limit their ability to effectively deliver their commitments.

o Rebadging may be a good thing to pursue; but it can also create liabilities for the client and issues for the affected employees. Go into it with caution.

Notification and Timing:

The Working Team (WT) in conjunction with the Executive Steering Team (EST) needs to develop messages for the enterprise, and the IT organization regarding the outsourcing initiative. There should be a series of scheduled announcements with the first being an update from the EST to the CEO and key staff. This message includes an overview of the project, an estimated magnitude in terms of potential benefits, number of impacted employees, and general project timelines. Depending on the firm's financial policy and the size of the contract, approval may also be required from the Board of Directors.

The CIO and/or the WT should meet with specific business partners that will be directly impacted by the project and review the same basic topics that were posted to the CEO. A general announcement to the IT organization should be made by the CIO explaining the vision for outsourcing including the specific business problem it will solve and introduce members of the WT. It is okay to mention that the outsourcing initiative will change the internal structure of the IT organization as well as telling the group that will there will be some new positions created while others will be eliminated. No need to go into specifics such as how many people will lose their jobs (or when) because that information will not be finalized for a few months.

Confidentiality is key, as the names will constantly change given ongoing dynamics inherent in projects. Even in poor economic times, people move from one job to another. Someone on the impact list might fill an open position or move to a different job. When this happens it doesn't make sense to backfill the position with another internal employee. Depending upon how far along the project is, there are some options to consider. If the supplier has been selected, one option would be to use a supplier resource to backfill the open position. This could be done using staff augmentation until implementation is complete. If a supplier has not been selected, perhaps the incumbent supplier (if there is one) could be used to backfill the position. The worst option would be to hire an internal employee to fill the position as they may soon find themselves on the impact list.

The termination date is a key element needed for notification. You can't have a conversation with an employee and say "your job is being eliminated and we'll get back to you when we figure out the date." Once you tell them their job is being outsourced, their focus will immediately switch to survival – where will I work, how will I provide for my family, etc? Determining the termination date for the impacted employee takes some thought and requires very detailed transition plans. As an example, if Suzie is doing "job X," Suzie will need to stay active until "job X" is successfully transitioned to the outsourcer. The transition process from start-up to steady state can take 4-5 months and could experience delays that may extend it. There will also be many others performing jobs that will also require termination dates.

Before the termination dates can be determined, a supplier partner needs to be selected and engaged. The Working Team along with the supplier needs to generate the transition timetable. My advice is to plan and budget to keep the impacted staff on the payroll for 4-6 weeks after steady state is complete. This gives the client the option to delay the announcement until the implementation process is well along its way. It also builds in a little cushion in case the project is delayed. Another element that comes into play is the minimum time required for notification. Giving the employee 60-days notice is common which means that termination cannot occur for 60 days from the time the employee is notified. The optimal time to announce the workforce reduction may be a few weeks before steady state is complete. By this time much of the knowledge has been transitioned and people are expecting some sort of notification anyway. Nothing will be gained by a premature leak of the names.

Short Term Organizational Reaction:

Few organizations react to outsourcing with rousing applause. Nobody likes change and large-scale outsourcing drives lots of it! As we are learning, there are a number of dynamic events at play when moving work to a managed services supplier. Keep in mind, the EST and WT have had ample time to absorb and become comfortable with the outsourcing idea – the rest of the organization has not. Business partner staff, although not usually impacted by layoffs associated with IT outsourcing, will be concerned with service degradation. Time should be spent with them to help assure their concerns are heard and valid issues addressed.

There are two completely different messages that need to be delivered regarding the outsourcing initiative. The first should post the enterprise that IT will be undergoing organizational change. As we noted earlier, tell them some new jobs will be created while others may be eliminated. Outsourcing is hard to keep under a bushel and this announcement will help explain the activities people will see around them. Also, many of the existing staff will be involved in the effort. The initial reaction from the IT rank and file will be one of survival. They will be obsessed wondering if their job is going to be eliminated. It is important to keep them focused and working on the objectives in front of them. The second message informs each impacted employee of his or her fate. The time between the initial announcements and notifying specific individuals can be a few months. Realize that anxiety levels will be rather high during this time. When it is time to deliver the news of job eliminations, it should be done over a very short time period. Don't drag this out over days and weeks, people will be thinking of nothing else anyway.

Long Term Organizational Reaction:

A positive benefit from outsourcing is the transformation within the IT organization, as it morphs to being "process driven" instead of "people driven." Well-defined processes combined with captured knowledge replace the reliance on that single individual who performed the job over many years. Let me explain this further:

➢ In the traditional IT organization, a person (lets call him John) will learn a great deal about the applications they work with every day. Over time, this acquired knowledge makes him a subject-matter expert. John enjoys the glory but rarely has the time or interest to document his knowledge or share it with others. The typical IT organization in the U.S. does not spend the time or money to train and backstop John either. Eventually John is soon labeled "irreplaceable." This new label restricts him from expanding his experiences and literally stops his career growth in its tracks. His manager will try to block him from being promoted to any other job since "nobody can ever replace him." I have seen too many cases where great employees are held back from a job move for months and in some cases are stopped from moving. This creates a devastatingly negative attitude - nobody wins. John will leave (one way or another) and when he goes, so does his knowledge. John's replacement has a steep learning curve to climb without a clue of how or where to start. Sadly, HR generally fails when it comes to eliminating the conditions that lead to "irreplaceable" employees and the CIO is oblivious to it because his direct staff propagates it. How many "John's" do you have in your organization.

Outsourcing to a managed services partner will provide the opportunity to change the internal workings of IT to become more process oriented. Well-defined procedures and the capture of explicit data form the foundation of the process-oriented organization. What is typically captured is the documentation of all code changes; records of how incidents were resolved; context diagrams; application maps and explicit data. If Arun is the resource replacing Jimmy, Arun will be responsible for filling in the gaps in documentation while categorizing as much application knowledge as he can. He studies past incidents to learn how they were resolved. After a few weeks, Arun ready to replace Jimmy and he will actually be tested to make sure he is qualified. Outsourcing partners will prepare back-ups for key positions using the recorded knowledge to do so. When it is time to replace Arun the transition goes as smoothly as possible. With a solid foundation of good process and recorded knowledge, the "process driven" organization is able to diminish the dependence on a specific person.

There will be a small number of employees needed to govern the supplier once the implementation is complete. The retained staff will be required to wear many hats. Using the process tools described above, the retained staff should be cross-trained in multiple areas. As their knowledge base increases, so does their value. They will soon be viewed as prime candidates for openings in other parts of the IT organization. When they are promoted and leave, others that have been cross-trained can fill the void and the "process" grooms the next replacement. IT folks, just like everyone else, want to maximize their potential. An organization that is process-oriented will create highly desirable employees with clear career paths that will be sought after by others within IT. This is a good thing because it strengthens IT, reduces the impact of turnover, and improves morale.

The *Inside* of *Outsourcing*

Chapter 10:

Supplier Relationship Management (SRM)

Supplier Relationship Management (SRM), as it pertains to IT services, will address the firm's overall use of third party providers. Table 10-1 highlights the SRM topics in this chapter.

Table 10-1 Supplier Relationship Management (SRM)

Strategic Concerns	Tactical Focus
Objectives of SRM	- Segment the suppliers - Continuous Improvement (CI) and supplier collaboration - Consistent, fair and ethical sourcing practices - Touch-point for IT management and Legal Services - Optimize Rate Cards and pricing via competitive bidding - Common escalation processes
Develop a Global Strategy	- Understand costs to assure apples-to-apples comparison - External vs internal roles - Leverage the market with supplier strategy by category - Negotiations with supplier and follow-up
Supplier Optimization	- Resolve common issues - Supplier performance - Training the organization - Change Management - Consistent SoW processes
Generate a Balanced SRM Scorecard	- Customer experience - Business process improvement - Innovation. Learning and financial benefit

As mentioned earlier, IT organizations need labor in order to function. That labor could be employees or third party providers. Chapters 3 & 4 discussed the three primary ways third party resources are used in IT: consultants; staff augmentation and managed services. The term "outsourcing" is often used as a reference to all third party relationships. However, some reserve "outsourcing" to only be used as a reference to a managed services relationship moving specific work and associated responsibilities to a third party provider. It is quite common that top tier suppliers are capable of delivering all three relationship forms.

Firms will benefit with a function dedicated to the procurement of IT services. As the use of third party, and especially outsourcing, has intensified over the past few years, the "go to market" strategy to procure IT services has evolved in a rather "helter-skelter" manner for some firms. In many cases, each IT manager created their own unique supplier relationships. Although the decentralized approach was well intended, the result has left IT organizations burdened with multiple sourcing processes, varied contract terms, suboptimal rates and a mix of disparate performance metrics.

The primary objectives of the SRM function will address the many issues related to the use of third party IT service providers. The very first item on the SRM agenda should be a survey of the IT client group to fully understand their concerns, frustrations, and issues associated with sourcing. Gaining an understanding of what is working well is important too. In order to address these concerns, the core focus of SRM should assure the following:

Supplier segmentation occurs along two dimensions – strategic and transactional.

➢ Transactional suppliers are typically used to manage specific transactions or deliver defined skills such as Java coding. The engagement is for a defined time frame and the supplier is engaged in a staff augmentation relationship.

➢ Strategic suppliers provide critical capabilities or "outcomes" for the organization.

   o Consulting services have many forms from redesigning the organization, to helping with strategic implementations. These are for a specific time frame with defined results.

   o Long Term Agreements (LTA) to provide Managed Services are only drafted with suppliers worthy of the efforts required to create and maintain a collaborative relationship. This enables successful outsourcing!

Continuous Improvement (CI) resulting in quality improvement, cost reduction, etc. must be committed to by the supplier and demonstrated on a year-over-year basis. The specifics of the CI plan should be established between the client and supplier at least annually, preferably a semiannual pacing.

Collaboration with suppliers includes bilateral sharing of goals, organizational strategies, and tactical plans for achieving them. Collaboration can include financial incentives or "gain sharing" opportunities for the supplier if they are significantly exceeded. Collaboration has the potential to take the relationship to its highest practical level. Below are critical steps required for collaboration.

➢ Internal Stakeholder Alignment; Supplier Selection; Contract with the Supplier.

➢ Set Governance Structure and Meeting Cadence.

➢ Use Change Management to Instill Collaboration.

➢ Measure Success.

Consistent sourcing processes need to be established and enforced across the organization. Suppliers want the client to be successful with outsourcing as it generates future business. They will be happy to help establish consistent processes, however, the processes they install will be "supplier-centric" based on their way of doing things. As the approved supplier list grows, the client finds itself needing to adapt to the unique processes within each supplier's internal structures. This creates confusion for the client. By implementing consistent sourcing processes across the client's IT organization, the tables are turned (so to speak) requiring each supplier to conform to the needs of the IT organization instead of IT conforming to the needs of each individual supplier. Examples of consistent sourcing processes include:

➢ Standardized RFP's and supplier evaluations.

➢ Standardized the SoW format to speed the approval process.

➢ Establishing SRM as the communication conduit for suppliers.

➢ Creating standard metrics to monitor supplier performance.

Fair and ethical practices create a strong foundation for both sides of the relationship. The two boundaries where money is exchanged in any business are the point of sale, and at the point of purchase. There are constant pressures to increase sales volume and decrease the cost to procure them. The buyer is trying to pay less while the seller is all about increasing revenues. The boundary between buyer and seller is ripe with opportunity for problems and when issues occur, it is usually along this boundary. When an employee of the buyer or the seller has a strong financial motivation to "seal the deal," some sort of ethical violation might be lurking in the shadows. In some cases tax laws may be violated. In order to mitigate any wrongdoing, firms publish a Code of Conduct addressing ethical standards and practices focused on reducing issues at the boundary. Employees are required to review and agree to follow the Code of Conduct.

Most violations that occur are blatant and easy to recognize, however, some may be less obvious and often seem rather innocuous...like these:

> The supplier offers an immediate incentive or "kick-back" directly to the client employee responsible for making a purchase decision. It may be a reward for actual or potential business. It can take the form of cash, lavish dinners, event tickets, paid travel, etc. This is an obvious case of an employee benefiting from the interaction. It is possible that their decisions are tainted with the lure of self-interest.

> The client agrees to pay now for services to be delivered at some future date. This usually occurs when projects have unspent money left over. Rather than give the project money back to a general fund, the client strikes a deal with the supplier to buy consulting hours that will be used in some future year. Although this may seem like a good decision, there may be revenue recognition issues involved posing a problem for the seller and the buyer.

> The client asks the supplier to defer billing for actual services delivered until some future time period to help manage budget problems. This seems like a good way to manage "non-budgeted" transition fees. Let's assume transition costs $500,000 and was not included in the department budget. A common practice is for the supplier to essentially "loan" the client the $500,000 and amortizes the repayment over the life of the contract. This is one of those gray areas that should be reviewed and approved by the finance and legal folks to assure things are set up properly.

One of my favorite stories I heard along the way is about an employee who we will call Tom. Tom was invited to play in a supplier-sponsored golf outing with all expenses paid. It was a wonderful event with great prizes, celebrities and lavish food and drink. Tom was a good golfer. He was so good that he actually shot a hole-in-one winning a brand new car. Once word got back to the office, Tom was told he had violated the company code of conduct by participating in the event in the first place. He was then told he would be terminated if he kept the car he had won for his hole-in-one. His choices were to return it, donate it to charity or face termination. He decided to donate it to charity. That was a good move for Tom. Unfortunately, the car was titled to him before he drove it home from the outing, thereby establishing him as the owner. The value of the car was considered income so he had to pay a sizable tax based on the $21,000 sticker price for the new car. When he donated it, it was valued as a "used" car losing about a 1/3 of the sticker price in the exchange. In essence, he paid tax on $21,000 in winnings but could only claim a $14,000 deduction. Hard to believe a hole-in-one could cause such problems – I am happy that my personal level of play will protect me from ever having to deal with that problem!!

A common "incubator" for violations is fraternization between the supplier and the client. A solid working relationship between the supplier and client is essential for successful outsourcing, however it can go a bit too far. I call this an incubator because it sets a tone in which the client (and the supplier) may take actions because of their personal relationship rather than their business relationship. Danger signs include:

➢ Frequent "pricey" dinners where supplier picks up the tab.

➢ Socializing with the supplier during non-business hours.

➢ Accepting all-expenses-paid invitations to supplier meetings.

➢ Agreeing to award future work without a formal RFP.

➢ Making decisions that do not benefit the client's firm.

Making SRM responsible for approving all RFP's, SoW's, rate cards, etc., establishes an "arms length" buffer between the client and supplier mitigating exposure to the firm.

Establishing SRM as a common touch point for legal assures that Master Services Agreements (MSA's) are set up with standard language for each supplier. The SRM team takes the lead working with suppliers. They are the liaison between the suppliers and their internal legal teams. Legal works with SRM to assure the MSA and subsequent Statements of Work (SoW) contain common terms and conditions protecting the interest of the client.

➢ The MSA is a contract that outlines the general legal relationship between the supplier and the client firm.

➢ The SoW is a contract between the supplier and client that outlines actions, timelines, milestones and deliverables for a specific project. The SoW will be a subcontract of and subject to the terms and conditions outlined in the MSA. We will go into more detail on the MSA and SoW in Section 3 of this book.

A close working relationship between SRM and Legal doesn't end when the contract is signed. There will be an endless list of questions that arise requiring a legal perspective during the life of the contract. A well-written MSA should outline, to the extent possible, how changes to the contract will be implemented. Legal should "sign-off" whenever changes are needed. Every so often, the supplier and client will not agree on how to interpret certain language within the contract. It is important to resolve these questions when they arise being careful to not use the contract as a club to beat the suppliers. I have worked with corporate lawyers well versed in how to engage contractually with IT services suppliers in a positive manner – they are worth their weight in gold. I have also worked with others who belabor every point in the contract and continually refer back to the details every single time something comes up during the engagement. This type of behavior can easily derail the whole relationship with the supplier.

A common touch point for IT Managers will make it easier for them to secure their sourcing needs. SRM must be designed as a service and avoid becoming a roadblock.

Templates and procedures should be core components of the service offering. They should be designed to help IT management work through the sourcing process quickly and effectively.

SRM minimizes collusion between IT managers and suppliers. Staff is works with IT managers on sourcing thereby minimizing the opportunities for collusion between the IT manager and suppliers.

As long as business needs are being met, the IT Manager should not care which service provider supplies the resources. Therefore, it is important to take the time to define those needs.

In addition to the technical skills required, there may be strategic concerns that need to be addressed. For example, assuring that only one supplier is used across a predefined application domain may be preferred. One supplier working across Supply Chain projects greatly improves effectiveness. Within any major application domain, a functional depth of knowledge is required as well as an understanding of how the applications share data horizontally across their boundaries. One supplier can manage this domain much more efficiently than many; this is especially true for Application Support.

It goes without saying that the needs of IT must compliment the overall needs of the firm. SRM balances objectives to achieve the best cost with the most capable suppliers. A common SRM strategy is establishing a limited number of approved suppliers for various types of work such as project management, infrastructure support, application support, etc.

Managing resource demand benefits both the client and the supplier. SRM needs to maintain a close working relationship with

IT management in order to anticipate future needs for third party services. They are responsible for sharing requirements with suppliers in order to assure they have the ability to respond to potential changes in resource demand.

Rate cards reflect actual hourly rates to be paid for various skill sets by location. The rate card is part of the MSA and is referenced in each subsequent SoW.

➢ For example, a Java coder with five to seven years of experience working in Bangalore, India may be billed at $30/hr while the same skill type working in Chicago may cost $75 per hour.

➢ Volume discounts, or rebates, are very common and should also be defined in the MSA. The size of the rebate should reflect total global volume with a supplier – the culmination of dollars spent across all SoW's. For example, if total volume of all SoW's with Supplier A reaches $10MM dollars in a given year, the rebate might be 3%. If the volume reaches $15MM, the rebate increases to 5%; at $20MM the rebate increases to 7% - you get the picture. The more the client spends with a supplier, the bigger the rebate. This type of discount based on volume "slabs" is the easiest to administer for both parties.

o Other types of discount schemes such as "promissory discounts" can be difficult to manage. An example: "the supplier agrees to pay a $100,000 rebate if they are used for 60% of all development work with the client". It sounds innocent enough, but the devil is in the details. First of all, the client and supplier need to agree on the unit of measure; is it 60% of resource hours or dollars? Does it include client resources, or just third party? Also, the client needs to "open their books" to prove exactly how much "development work" there was – this is not easy. Do minor code change efforts done by support staff affect the total? What about money paid to software vendors to help implement a new release of their software package?

Consistent payment terms and processes lead to internal efficiency for the IT organization. These too should be defined in the MSA and apply to all SoW's. Most suppliers want to be paid as fast as possible; clients on the other hand want to delay payments as long as possible. Most agreements state the supplier will be paid by some date (35-50 days) from receipt of invoice and carry a penalty if paid late. Understand "how" payments will be processed. In order to pay the supplier, the payable system requires an invoice. The invoice requires a PO (Purchase Order). The approval chain can include multiple steps depending on the size of the payment and approval levels established for your firm. So, plan ahead and don't wait until the last minute to start the process.

➢ One of my clients had a 5 year, $60 million dollar contract which, because of its size, required approval by the Board of Directors. To manage the budget, the contract was broken into 5 individual PO's (one per year) each worth $12 Million. The CEO was required to sign-off on each PO's. The supplier then submitted a monthly invoice of approximately $1MM which referenced the PO.

Competitive bidding is a good practice to achieve the best value. All work performed by third party suppliers should be bid on a competitive basis to assure best price, minimize collusion and eliminate shady activities.

➢ The bidding process should start with the Request for Proposal (RFP) and end with an accepted bid - RFP's will be explained in a future chapter.

➢ SRM creates a "go to market" strategy. This starts with the creation of short list of suppliers for large domains of future work. Creating an RFP for every SoW is not an effective use of time. RFP's should be launched for each of the large domains. For example, there may be an RFP launched to source a supplier to work on all future IT projects in the Sales Systems domain.

➢ The winning bidder would be awarded the work in that domain for some time frame. This should not be construed as a financial commitment. It is nearly impossible to determine project spending for 5 years into the future. This is a process to establish the preferred supplier for the domain. Each future project for the preferred supplier would require a SoW to define objectives, milestones and establish the financial commitment.

A defined escalation process is important to assure that issues between clients and suppliers can be resolved quickly. When engaging with the outsourcing supplier, "key roles" should be established. Client and supplier leaders should be named for each role. These leaders must have both responsibility and authority to not only deliver key aspects of the transition but to resolve issues at their level whenever possible. This will expedite resolution and build a stronger working relationship for the team. Examples:

➢ Executive Leadership: Responsible for the overall initiative.

➢ Contract Management: Develop the contract with legal.

➢ Engagement Staffing: Name SME's and supplier resources.

➢ Process Integration: Seamlessly merge mutual processes.

➤ Logistics: Plan for workstations; user ID's; space planning; etc.

➤ Infrastructure: Offshore connection; bandwidth; equipment.

➤ Communications: Internal/external PR and announcements.

➤ Transition Leaders: Manage migration to managed services.

➤ Program Managers: Responsible for application clusters.

➤ Cluster Leads: Manages a specific suite of applications.

Common formats for Statements of Work (SoW's) will expedite the internal approval process, as they will contain required legalese and inclusions. SoW's are contracts for performance and should contain common SLA's, standards and protections across all suppliers. There should be two general forms for SoW's; one designed for staff augmentation work and the other for managed services.

Staff augmentation SoW's typically focus on skill requirements. For example: "We need 2 Java coders for 2 months." This becomes the basis for the SoW. Essentially it addresses the "inputs" (Java coding) required for a client-managed project.

Managed Services SoW's are different and identify the desired "outcomes" resulting from the work effort. For example: "We need a team to manage break-fix and maintenance activities for Warehouse applications." Or "We need a team to deliver an application linking shipments to our carrier partners."

➤ Obviously there are more explanatory comments required, but the gist of the idea is to write the SoW reflecting the outcome of the work process rather than the inputs needed. The supplier will be responsible for providing the inputs required to deliver the defined outcome.

The global sourcing strategy should be one of the primary outputs of the SRM team. The sourcing strategy is an important tool identifying approved supplier partners and defining the procedures for IT to follow. This work should be a joint effort between SRM and the IT managers that will be using this service.

Understanding base cost is critical when determining the benefit may be gained by the outsourcing project. We covered this topic in great detail in Section 1 as we built the business case. It is equally important to understand the specific services the supplier will deliver and how they impact the base cost as well.

Apples-to-apples bids simply don't exist in nature. Suppliers will present their data in a slightly different way. Creating a bid template that attempts to align feedback from multiple suppliers for comparison will be helpful. NEVER EVER accept the data on the template without first understanding what is driving every number on the page. A thorough understanding before the contract is signed will minimize confusion down the road.

Creating the strategy starts by asking IT leadership to define roles for internal employees and those to be outsourced. Examples:

➢ Application Support for credit card transaction systems, legal files, closely guarded formulation data and other sensitive areas may likely be assigned to internal employees to protect the firm's Intellectual Property.

➢ For Development projects, internal resources may be needed to gather requirements and weigh in on the architectural design and final solution. Third resources could write specs, cut code, test and install the finished product.

➢ Infrastructure Support and maintenance is a large domain that firms often outsource. It is wise to limit the number of eggs in one basket. The sheer size of this domain would suggest that the chosen supplier would only work in this area.

The marketplace survey is conducted after broad domains of work have been determined. SRM identifies the best way to leverage available suppliers for each category. Ideally, there would be 3-4 suppliers asked to prepare an RFP for each domain. It may make sense to have more than one supplier for a specific domain. At the same time, a supplier may be assigned to more than one domain. A watchful eye must be kept on the number of eggs in the basket.

SRM leads negotiations with suppliers; however, the IT manager responsible for the work needs to be closely involved as well. Supplier negotiations follow a lifecycle. The early stages of the lifecycle involve aspects about the general service delivery, staffing/pricing, risk assessments, etc. In later stages, specifics about the work to be performed are discussed. In the final stages, the supplier is performing the work and meeting SLA's. SRM is more heavily involved in the early stages and the IT manager in the later. As the project moves along, IT management takes control and SRM should only be involved if serious delivery failures occur.

There is a need for frequent "top-to-top" meetings between SRM leadership, IT management and the various suppliers. As the marketplace and service offerings continually evolve, SRM needs to stay on top of this to assure the organization can gain additional value from new services the supplier may offer.

Supplier optimization is the result of ongoing efforts and relationship improvements with the supplier. SRM has four major areas that if done well, will result in a highly effective relationship.

Resolving common issues that a specific supplier is exhibiting across multiple projects requires SRM involvement.

➢ An example may be that a supplier not following the defined process steps for invoice creation.

➢ Another may be a supplier with a tendency to shortcut testing processes.

Monitoring supplier performance at the project level is the responsibility of the IT project manager. However, creating the tools to monitor performance such as Service Level Agreements (SLA's) and/or Key Performance Indicators (KPI's) fits well within the role of SRM. SLA's and KPI's are the backbone of every successful engagement and therefore must be worded properly to incent and reward desired actions by the supplier. This takes a certain level of skill and centralizing the role within SRM assures consistent use across all projects.

➢ Note: The major difference between SLA's and KPI's is that SLA's are bound by the contract and may carry defined financial penalties for the supplier when missed. KPI's are simply data trends of a specific performance factor and usually don't carry any associated penalties. KPI's may become the basis for new SLA's to be added to the contract with approval from both parties.

Training the organization & change management. Executive edict is one way to force adoption of any change - and outsourcing represents a huge change. However, the better way is for SRM to actually train the organization on the services provided and their value to the organization. Transformation to a supplier-based environment will require a great deal of effort. Once processes are defined and procedures are in place, SRM needs to create curricula to educate the organization on how to use them.

➢ I have found a combination of smaller "brown-bag" sessions and larger general meetings work well to deliver the message and train the organization.

   o The general department meeting is a good place to share the reasoning behind the SRM function. Describing service offerings and how they will improve the internal workings of IT should be the main emphasis. The general meeting is a good place to address questions from the broad organization. For the first year, progress postings and success stories are good topics to include at future department meetings.

o The brown-bag sessions should be designed as a series of meetings that cover a number of topics. They should be short, maybe 30-45 minutes and are usually scheduled over the lunch hour (hence the term "brown bag"). Their focus should be on new procedures the IT staff will be using to interact with SRM and suppliers. These sessions should be advertised and repeated a couple of times to assure the maximum reach to IT. If possible, they may be offered as internal "webinars" as well.

Consistent SoW processes not only improves efficiency in the IT organization, it provides a common format for the suppliers to use as well. When the specifics required for a rapid approval are known up front, it allows the supplier more time to focus on content of the SoW.

Creating a balanced scorecard is an important output measuring how well SRM is serving the often-divergent needs of the organization. These include providing the skills needed by IT management, client/user satisfaction with IT in general, ongoing IT innovation and of course improved financial results. Measuring results periodically will show how they are trending over time.

An improved customer experience is the hallmark of a well-functioning SRM team. This is usually based on developing repeatable processes that the IT customer can easily interact with. The end result will give IT the ability to consistently attain resources with acceptable skills at a competitive cost.

Business and process improvement come from a data-oriented approach to problem analysis. Top tier suppliers are dedicated to identifying process improvements and making the minor enhancements required to achieve them. Capturing statistics on minor enhancements and sharing them with the Business Partners on a periodic basis is a great thing to do.

Innovation and learning are desired characteristics in any organization. SRM is in a great position to capture "best practices" from various suppliers and introduce them to the organization.

Financial benefits are easy to track. It is the single metric that management will use to determine how well SRM is functioning. The ability to track year-over-year savings as well as Return on Investment (ROI) has to be a major component of the SRM strategy.

One last topic before concluding this chapter is the best organizational fit for the SRM function. Should SRM be "owned" by IT or Procurement? Without customers, an organization will fail. Suppliers are integral to success as well. The importance of managing customer relationships is well recognized by many firms with strategies and processes to manage it. Customer relationship management (CRM) software is quite mature and is offered by a broad range of vendors. Until recently, however, Supplier Relationship Management (SRM) has been the forgotten frontier.

According to Alan Day, "it has been estimated that SRM can deliver estimated savings of 3-5 percent. More important is the fact that a lack of focus on SRM presents opportunities for suppliers to exploit the organization using a "divide and conquer" tactic across the various IT functions." Effective SRM focuses on new-product development; better customer experience; improved organizational capability and reduced total cost of ownership. Strategic suppliers to IT operations should be managed under the SRM umbrella. Examples of strategic suppliers are those with whom the IT team spends significant amounts of money, those that provide critical products or services, and those with a track record of providing innovation.

Let's look a little closer at the question of which broader corporate function should own the Supplier Relationship Management team. Defining "ownership", what it means in practice and where it lies organizationally can be problematic. A broad range of ownership structures exist within and between the various functions and their suppliers. Here are a few examples:

➢ Budgetary ownership: "Owning" a budget often brings with it the right to decide who to spend the money with and how to manage the relationship. When the function (rather than procurement) controls the budget to buy IT services, they naturally feel they should have a major role in supplier selection and ongoing supplier reviews.

➢ Technical ownership: The IT function designing a new application will often dictate which suppliers should be used. This of course can be manipulated by aligning the technical specifications to a specific supplier's product or service.

➢ Strategic ownership: The executive team believes supplier relationships are critical to the ongoing success of the firm and feel they are best positioned to manage them.

➢ Financial ownership: Since it is responsible for processing invoices and guarding cash flow, finance might lay claim to owning relationships with particular suppliers. However, when the score is tied late in the game do you put the scorekeeper on the field to win the game??

➢ Operational ownership: People who deal with suppliers on a daily basis feel that they are best positioned to manage them.

➢ Legal ownership: Because they are the guardians of contracts, legal believes it is best positioned to manage suppliers. I'll ask again, when the score is tied late in the game, do you put the scorekeeper on the field to win the game??

➢ Governance ownership: The nature of suppliers' operations or goods and services lend them to being managed by an area familiar with the business processes and technologies associated with the supplier. For example, the packaging engineering team may manage a supplier of corrugated materials.

Ongoing viability ownership: Which organization is best suited to fill any future openings in the SRM group with capable managers? Is it easier to groom technology-savvy procurement specialists or procurement-savvy technologists?

An argument can be made that specialized purchases such as IT services and applications are better dealt with by the relevant business functions. That is where the highest level of expertise (a.k.a. technical ownership) resides. Assuming this is the case, does it necessarily follow that they should also own the overall supplier relationship? There are a number of specific skills and knowledge required to optimize the relationship with suppliers that may not be the forte of the individual functions. These skills, however, are often core within the procurement function and include:

➢ The ability to accurately model costs.

➢ Comparing suppliers on an apples-to-apples basis.

➢ Utilization and coordination of cross-organizational teams.

➢ Reducing the impact of price fluctuations on cost structures.

➢ Early supplier involvement in product/service development.

➢ Dissemination of knowledge across the firm's value.

➢ Service planning and design synergy.

➢ Use metrics to drive and manage change for both organizations.

➢ Improved risk-management and continuity of supply.

➢ Access to, and speed of, innovation.

One can easily posit the argument that SRM itself is a specialized process best managed by Procurement.

The answer to the question "who should own" SRM can be debated long after the cows come home. I believe that no single person or specific function should "own" the entire supplier relationship. Success requires that two main elements work hand-in-hand:

➢ Creating a structured "deal" with the supplier.

➢ Strong, active implementation from stakeholder functions.

Supplier management works best with a structured deal. Therefore someone needs to be responsible for "creating the deal" with suppliers. There is also the need for the deal to be implemented and managed on a day-to-day basis. "Ownership" of these processes (creating the deal and managing the work) depends a lot on the strength of talent in IT as well as in Procurement. The answer to whether IT or Procurement should lead depends on the relative strengths of each organization.

To effectively lead SRM, Procurement must be seen as an executive decision-maker with responsibility to create and govern processes to manage supplier relationships and disseminate information across the enterprise. This structure creates an "honest broker" detachment from daily operational pressures to minimize collusion and inappropriate activity. If this structure is chosen, the functions will still own the operational day-to-day relationship in accordance with their specialization. If/when these start to fail and affect the supplier relationship they would be escalated to Procurement to facilitate resolution. Conversely, IT may be seen as the stronger organization and should take the lead on SRM. Either way, SRM being a relatively new function must tread lightly through politics surrounding ownership that will pose interesting challenges.

A Procurement group lacking a depth of technical knowledge will have a difficult time re-positioning the "ownership" of an existing supplier away from an established organization such at IT. This is especially true if that organization holds a stronger position within the firm as well as with the supplier.

➢ Will the CPO be willing to take on the CIO for ownership of a large technology supplier?

➢ If SRM is managed by Procurement, it should create supplier management frameworks, processes and governance. This will allow procurement to influence how the broader organization interacts with suppliers and ensure a consistent approach.

➢ If Procurement leads, it must position itself as a "central clearing house" for information flow, allowing visibility on how the supplier is being managed at an operational level by the other departments. Caution should be taken to avoid the potential for this positioning to be seen either as a purely administrative task or, worse yet, a bottleneck.

For companies within which IT has the relationships with technology suppliers, it will be a challenge for Procurement to take the lead. The CPO needs to create both a compelling case for SRM ownership within procurement, and create a method for achieving and delivering the benefits. Procurement will need to staff SRM with IT savvy resources to achieve its objectives – most of these reside in IT. A great deal of effort is required to segment the supplier base along the spectrum from transactional to strategic; align appropriate management models; gather, analyze and share key information; and create an environment in which supplier innovation can be captured and exploited ahead of the competition. Will IT willingly supply the resources for this to happen? If so, what is the career path for them? Will they come back to IT after a few years in the SRM group or will they move to higher levels of responsibility in Procurement? A well-defined career path will help prospective resources to see the move from IT to Procurement as a positive move for their future.

In conclusion, the answer to the question of SRM "ownership" is that it all depends on the specific firm in question and the history of how the two functions have been utilized.

➤ Procurement's orientation towards frameworks, process and governance is important when working with external suppliers.

➤ The IT function can best determine the supplier's capabilities, their work content, cost, quality and long term viability.

➤ Procurement is best suited to access supplier information given its existing relationships with suppliers across the organization.

➤ The functional organization knows technology and is better suited to foster an environment in which suppliers will share improvement ideas and innovation.

➤ Given the multiple levels of SRM ownership, procurement fulfills a useful role as an "honest broker" between the supplier and the wider IT department.

➤ Visibility to the broad needs of the organization coupled with knowledge of the supplier's capabilities allows procurement to make informed buying decisions.

In order for Procurement to effectively lead SRM, they need people with a strong technology background. If IT leads SRM, they need to exhibit procurement skills to assess and evaluate suppliers and generate contracts that deliver value to the firm. As mentioned earlier, the question of SRM ownership all depends on the relative strengths and capabilities of these two organizations. There is no right answer that fits every organization.

This concludes Section 2 on Organizations Readiness. Our next section covers Supplier Selection beginning with the Request for Proposal (RFP), followed by chapters on the Master Services Agreement, Statements of Work, and Operating Agreements such as Service Level Agreements and Operating Level Agreements. The final chapter in Section 3 will walk through the due diligence process.

The *Inside* of *Outsourcing*

Section 3:

Supplier Selection

Section 1 focused on creating the business case and Section 2 on preparing the organization for outsourcing. I know it is painfully obvious, but there can be no outsourcing until a supplier is selected. I have had the honor of working with (and learning from) some excellent sourcing experts. Choosing the supplier is a critical step in the outsourcing process and should not be taken lightly – you win or lose by the supplier you choose!

Section 3 describes processes used to select your IT partner. Chapter 11 will focus on the supplier interviews and Request for Proposal (RFP). The Master Services Agreement (MSA) will be covered next which describes how the two parties will interact with each other and touch on common inclusions or Terms and Conditions known as "boiler plate" language. Chapter 13 goes into details about the Statement of Work (SoW) describing the work to be done, financial arrangements, and performance milestones. Operating metrics form the backbone of supplier partnerships - Service Level Agreements (SLA's) and Operating Level Agreements (OLA's) are addressed at length in Chapter 14. Due Diligence, or final review of the agreement before signing the contract is described in Chapter 15.

Areas of Focus: *"The* Inside *of* Outsourcing"

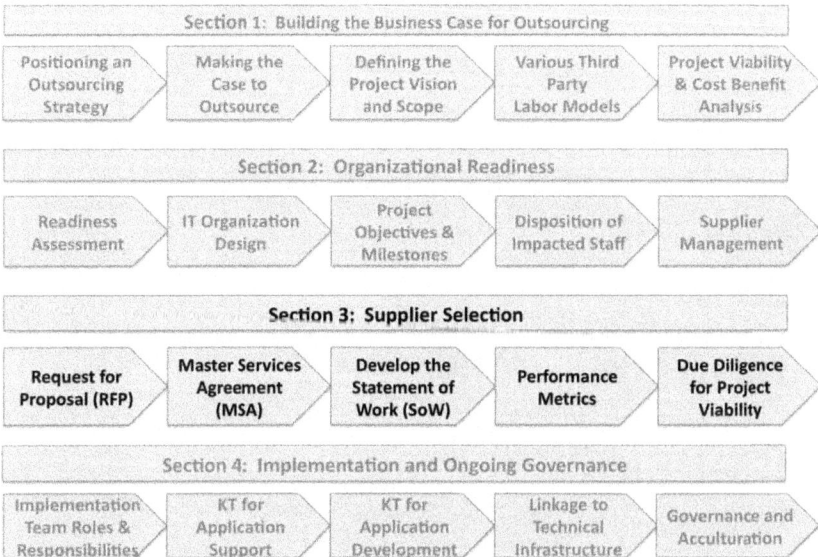

Section 1: Building the Business Case for Outsourcing				
Positioning an Outsourcing Strategy	Making the Case to Outsource	Defining the Project Vision and Scope	Various Third Party Labor Models	Project Viability & Cost Benefit Analysis

Section 2: Organizational Readiness				
Readiness Assessment	IT Organization Design	Project Objectives & Milestones	Disposition of Impacted Staff	Supplier Management

Section 3: Supplier Selection				
Request for Proposal (RFP)	Master Services Agreement (MSA)	Develop the Statement of Work (SoW)	Performance Metrics	Due Diligence for Project Viability

Section 4: Implementation and Ongoing Governance				
Implementation Team Roles & Responsibilities	KT for Application Support	KT for Application Development	Linkage to Technical Infrastructure	Governance and Acculturation

The *Inside* of *Outsourcing*

Chapter 11:

Supplier Interviews and Request for Proposal (RFP)

Table 11-1 highlights the steps involved in selecting a supplier partner. The selection process ultimately leads to the Request for Proposal (RFP), which is simply a formal way of asking a supplier to propose their solution to your outsourcing project.

Table 11-1 Supplier Interviews and Request for Proposal (RFP)

Strategic Concerns	Tactical Focus
RFP Objectives	- Create a list of guiding principals - Conduct supplier interviews - Invite suppliers to bid - Evaluate supplier responses - Award the work to supplier with best overall fit
Guiding Principals	- Supplier segmentation - Criteria for inclusion in the bidding process - Supplier concentration - Shared expectations
Supplier Interviews	- Describe the work in detail - Chance to see the supplier's advertisement for their firm - Talk about similar engagements the supplier has had - Visit supplier's facilities
Invite Suppliers to Bid	- Develop a common response template - Share and explain the bidding template - Share expectations - Send RFP to select suppliers
Evaluate RFP Responses	- Create the scoring template (it is only a guideline) - Review supplier responses - Narrow the search - Schedule a combined Q&A session
Award the bid	- Fine tune bids with finalists - Make final selection and begin contract negotiations

A well-designed RFP results in documented project requirements; a "fair" market-based approach to awarding work; and a cornerstone to consistent IT sourcing. It also shortens the turnaround time required to secure a valid bid from selected suppliers. The alternative to using a "process oriented" RFP approach would be to invite suppliers to come in and figure things out on their own, eventually gathering enough information to generate a proposal. Procurement can assist in identifying incumbent, developmental, niche and industry leaders in the marketplace.

The objective of a formal RFP process is to identify the best supplier for the work required. There are 5 major work segments that comprise the RFP process:

1. Create guiding principals to help focus the effort by eliminating random approaches to supplier selection.

2. Conduct interviews with potential suppliers to understand their responses and learn a little bit about their personality too.

3. Invite a few key suppliers (four or five) to review the RFP and bid on the work. It is important to provide a common RFP request to each supplier to ensure a fair and unbiased bidding process.

4. Evaluate supplier responses on an apples-to-apples basis assuring common terms and conditions; consistent scorecard comparisons; and knowledge of which suppliers offer value added service and continuous improvement. Award the bid to best overall supplier.

Guiding Principals will bring focus to the RFP effort by insuring key topics are addressed and working with suppliers that are best suited for the intended work.

Supplier segmentation is the process of looking at the field of potential IT partners and sorting them into different categories in order to identify the right suppliers to invite to bid on your project.

Some suppliers are best suited for application support activities, while others have practical expertise at managing networks and infrastructure while still others are have a strong track record in managing development projects. To be sure, some top tier suppliers can, and do, provide solutions across many categories but most focus on just one area. One should create a subset of the suppliers with the expertise to perform the work needed for the project in question. Procurement can assist in identifying new suppliers as well as the industry leaders in the marketplace.

The size of the supplier is another factor to consider when

segmenting the market – try to match suppliers to the potential engagement value. The natural order of things dictates that small suppliers fit better on small engagements while large suppliers fit better on large engagements. The reality of life is that clients in small companies will more than likely have small projects (relative to large companies) that are more attractive to the smaller outsourcing firms. Larger suppliers with market caps in the $25-$35 Billion range will not be interested in small efforts unless they lead to much larger engagement. For example, a project that generates $2 Million per year in revenue is small and not attractive to the top tier suppliers unless of course that $2 Million project has potential to grow to $15-$30 Million per year. The internal operating structure, resource base, and processes of a supplier are geared to match the anticipated revenue of the client. That small $2 Million per year engagement doesn't fit the large supplier as well as an engagement of $20 Million per year. Larger outsourcing firms have the infrastructure and resources to easily absorb and manage large projects. Small outsourcers on the other hand might have a difficult time staffing up the resources needed to provide the same level of service.

The supplier's geographic location should be an important part of your final selection criteria. Within the Asia Pacific (AP) region, India has a richest history and command of the English language and customs – remnants of their history in the British Empire. In the early 1900's the Philippines started teaching English due to the U.S. occupation. China and other AP countries are learning English and catching up. Outsourcing to Asia Pacific is a 10-12 hour offset from local U.S. time zones while Latin America may only be a 3-hour offset.

➢ The 10-12 hour offset may be great for application support but is not conducive to agile development strategies.

➢ Suppliers operating from India and supporting standard U.S. work hours may experience high turnover. The Indian resources would be working third shift to support the U.S. standard work hours.

It is also important to consider the impact a global multi-region contract will have on local markets. One size doesn't always fit all in this case. What may work well in the U.S. and Europe may not fit the needs of smaller markets in Asia Pacific and Latin America.

The supplier should agree to deliver performance metrics by market area. Global SLA reporting may miss local market issues. Assuming that the smaller markets are doing well just because results are good in the U.S. and Europe can be a big mistake. This could undermine the success of outsourcing in smaller markets.

The service package and cost structures must be modified to fit the smaller market areas. The business case and specific problems to be solved by outsourcing may not be consistent across all countries. Remember, identifying the problem to solve and creating a solid business case are the first steps to a successful project.

One of my clients had a difficult time implementing a major Support contract across the many small countries in Latin America. Unlike the U.S., each country has unique budgets, currencies, taxation and legal considerations. They simply couldn't afford to take on the burden the over-arching support contract presented.

Procurement should create a "ready list" of potential suppliers to choose from. These suppliers should be financially viable as well as proficient in the type of work they specialize in.

There are many criteria to consider when selecting suppliers. Some of these are quantitative while others are more qualitative such as:

➢ Does the supplier have a track record of proven ability? Have they done the type of work being defined in the RFP or will this be a learning experience for them…and you? If willing to incur the risk with using an unproven supplier, it is possible to negotiate better pricing – but tread cautiously. It is relatively easy to search for "success stories" and client testimonials on the top suppliers, but it may be a bit more difficult if using a smaller supplier. Ask them to provide a reference list of clients to interview. I have participated in a number of reference interviews and I am always impressed by the depth of questions potential new clients have, which is one of the reasons I decided to write "*The Inside of Outsourcing*."

➢ The supplier will be happy to share their experiences and expertise as well as the names of other clients on their "mantle piece." Each will be happy to tell how they can add value to your project. Prepare questions for the supplier to answer. This is your chance to learn about them so do your homework well and incorporate results into supplier evaluation scorecard.

➢ Personal history with the supplier can provide a good insight about how successful they may be on your project. Glowing praise is a good indicator but be prepared to listen to some negative feedback as well but listen with an open mind! One fact of life is that the supplier is always blamed for a failed project whether it is truly their fault or the fault of the client. Client teams are notorious for unrealistic expectations, poor requirements definition and not being prepared for outsourcing. These all work to the disadvantage of the supplier often leaving them with a black eye.

How much work should be awarded to a specific supplier? This is an important guiding principal that most client firms address. The old "too many eggs in one basket" idiom is often cited in order to keep the supplier from becoming too domineering. If one supplier has a monopoly on all IT services, one could argue there exists a risk of becoming a captive client. So what is the right level of concentration vs any trade-off in operational efficiency?

One approach is to segment the work by creating major domains within which one or at most two suppliers will be chosen to work. Application Support, Infrastructure Support and Application Development are three distinct domains.

As we touched on earlier, it makes sense from an efficiency perspective to select one supplier for application support work. The nature of support requires cross-application competence. From managing Support Teams, I have found that the application itself rarely breaks – problems usually occur at the interface layer between applications. For example, the individual apps within major Enterprise Resource Planning (ERP) solutions generally run without issue. Incidents occur when the core ERP applications exchange data with non-ERP applications that exist across the enterprise. Understanding horizontal flows across the interface and the actions required to keep them flowing is best suited to a one-supplier model.

A similar argument has been made for infrastructure support. It is more difficult to manage infrastructure when each technical environment has its own supplier. Success relies on common performance standards and metrics across all environments. These are difficult to achieve in a multivendor model.

Application development requires deep knowledge within a relatively narrow functional band. It is possible to create sub domains with a couple of different suppliers each being responsible for one or two domains. For example, application development within the Supply Chain function could be one domain while development within the Sales function might be another.

My suggestion is to select three to four suppliers for major IT areas. One covering application support, a different one for infrastructure support, and two or three others for application development across major functional domains. Each domain should go through a formal RFP process awarding the work to the best supplier for that domain. Be sure to keep the RFP process competitive across the supplier set in order to ensure the best quality, service and pricing is achieved.

Managing expectations is difficult yet rewarding. Bad things generally happen when expectations are not aligned. For example, if you go into your local fast-food restaurant expecting a 5-star meal, you will be disappointed every time. If you set your expectations to the world of probable outcomes, you may increase your chances of an expected experience. It is important that both client and supplier are on the same wavelength when it comes to expected work activities and how they will be performed and measured. The RFP must contain language that spells these out to minimize confusion down the road.

The primary output of the Guiding Principals process is a short list of suppliers, maybe four or five that fit the needs of the project in question. Once the short-list of suppliers has been determined it is time to set up interviews with each one.

Supplier Interviews give the client and supplier a chance to meet, exchange information and get a feel for how the relationship may progress. There are a number of common topics that should be covered with each supplier in as similar a format as possible. I suggest limiting the entire discussion to no more than three hours. Send an agenda to each supplier prior to the meeting so they can prepare. Allow time for the supplier to give a brief advertisement of their firm. This is also an opportunity for the supplier to conduct due-diligence and collect data in order to provide a valid bid.

Most of the meeting time should be spent discussing the details of the project. Be sure to put this in a format they can take back with them, as it will become the basis for their initial bid. The detailed RFP information documents should include:

➢ Description of the work the supplier will perform and outcomes expected from that work. Examples of "outcomes" include:

- o Supplier will manage support for all applications; maintain 98% uptime; resolve all Sev 1 tickets within 4 hours; deliver 300 hours (or 10% of staff effort) of minor enhancements each month.

- o Supplier will create an application showing inventory levels by item, by location and perpetuate availability based on forecasted demand and planned replenishments.

If there are any specific "use cases" or "scenarios" written, these could be provided as well, but it is more likely that they will arise in the requirements gathering phase after the supplier is selected and the project is launched.

Inform the suppliers about the timeline for the project. When will they need to return their bids, anticipated date for contract completion, date expected for the project to be completed.

Describe the application environment. Number of applications in scope; number of client resources (or other third party) currently performing the work; organization structure; business partner relationships and number of users; code development & testing methodologies; standards; documentation; security protocols; etc.

Overview of the technical environment: Servers; databases; packaged solutions; context diagrams; telecom; firewalls; etc.

Descriptions of the applications the supplier will be involved with: Packages or in-house developed; licensing; functionality; age; number of direct users; transaction volumes; platforms used; number of recorded incidents per time period; documentation; current issues; etc.

Allow time to discuss similar work efforts the supplier has been engaged in. To the extent possible, try to get a feel for the cost (or at least the number of resources used) to deliver that service. Discuss the type of skills required and which resources will be working remotely from offshore vs in the client's office.

Ask each supplier to explain his or her Knowledge Management process. How is it done? How long is each phase expected to take? Retention processes? Make sure your company maintains the keys to the kingdom (knowledge) to facilitate the transition to a new supplier in the future. Remember, your company (not the supplier) should own all past, present and future content from the Knowledge Management process.

Visiting the supplier's facilities is recommended before making the final selection. Most India-based suppliers are happy to host clients making the effort to travel to India for a visit.

➢ Most of the offshore offices I visited were very nice with manicured lawns, reflecting pools, and workstations with the latest technologies. The surrounding neighborhoods, on the other hand, were often very poor with dirt roads and cows roaming the streets – the contrast is striking.

➢ Ask to see the work areas and employee amenities offered. Look at training facilities and ask about Knowledge Transfer processes and tools. Check to make sure the work areas are secure. It may be a good idea to include someone from your security team on the visit to address confidentiality concerns.

➢ Make sure to have discussions about Disaster Recovery and Business Continuity Procedures to understand how the supplier will manage through a power shortage or other calamity. A major portion of the team will be offshore and when the power goes out, they stop working.

➤ If possible, meet with teams working on similar projects. Look for evidence of total quality management, lean, ITIL framework and process metrics such as run charts, Pareto diagrams, etc.

➤ Try to set up time with the key staff that may be assigned to work on your project. Building rapport is very important and time should be taken to do this whenever possible. Get a feel for the attitudes of the team. Team structure is a good review topic. Gain an understanding of how the supplier recruits talent, balances their career progress with continuity to clients, and turnover.

Remember, it takes about 16 hours of flying time to get to India from the U.S. so It is best to plan a multi-day trip and schedule as many supplier visits as possible during your stay. Many of the suppliers have offices in the same cities including Bangalore, Hyderabad, Chennai, Mumbai and Pune. Due to heavy traffic in these cities, I would suggest limiting supplier visits to no more than two in any one day.

CAUTION: The suppliers greet visitors with open arms, flowers, and of course a 7-course lunch. Regardless of the time of day, there will be a 7-course lunch. One must be extremely cautious and avoid eating raw or undercooked foods, and anything prepared with unsterilized water (including fresh fruit, tap water and ice) or risk the embarrassment suffered by one of my team members on a trip to Bangalore. The poor chap ate something that resulted in a severe intestinal rebellion. Unfortunately he was trapped on a bus the next day with his coworkers. Death, for him (and us), would have been an upgrade!!

➤ Most suppliers will help with travel planning details and also provide topic ideas to review during your visit.

➤ Once the supplier is selected, it is important for Working Team members to plan a visit to their facilities during the KT process as well. Annual meetings at the supplier's facilities will nurture the relationship ongoing.

➤ Travel to India is costly[5]. Plan on $10-$12,000 for a one-week stay per person to cover airfare, hotel and meals. Make sure to include travel costs as part of the project start-up budget and also in ongoing departmental budgets over the life of the relationship.

A formal RFP Response Template should be sent to those suppliers that pass the initial interview cycle. Some may excuse themselves from bidding once they have had a chance to better understand the

---

[5] *Travel costs based on 2011 prices.*

project details. I have found that if left on their own, suppliers will complete their response to the RFP in various forms and formats. The client then needs to recalibrate multiple responses in order to compare them on an apples-to-apples basis. Preparing and using an RFP response template will help keep the responses more uniform and easier to evaluate. Most clients will create their own template in order to achieve the following objectives:

➢ A common template used by all prospective suppliers will help focus the RFP responses to key questions making comparison fair and easier to score.

➢ Important elements to include on the template:

- o A recap of project details discussed during the interviews along with new information that may have surfaced.

- o Timeline by which supplier will deliver project elements.

- o The template will focus on the ongoing cost of resources by asking:

  - ▪ Number of resources by skill level required onshore and offshore and their hourly rates.

  - ▪ The annualized total cost per resource plus anticipated rate increases.

  - ▪ The number of work-hours per week for offshore and onshore.

  - ▪ Management level resources the supplier requires.

  - ▪ Any other ongoing costs such as connectivity for offshore resources.

  - ▪ A recap of total costs per year for 5 years – this will be a primary element of the cash flow analysis.

- o Commitment to annual cost reductions over the life of the contract or 5 years, whichever is greater.

- o One-time costs that may be incurred during Knowledge Transfer (KT):

  - ▪ Cost of both trainers and new resources being trained.

  - ▪ Cost of hardware required for offshore connectivity.

  - ▪ Travel between India and U.S. for those working on KT.

- o Definition of all critical SLA's, OLA's and KPI's the supplier is expected to deliver. Actual target levels, minimums and penalties will be refined in the final contract.

The RFP should also recap shared expectations between the supplier and client. Allow space on the RFP for the supplier to enter additional innovative ideas and services that may be of value to the client. Be sure to understand all cost inputs for both negotiation and validation purposes. Can the supplier perform for the price they have quoted?

Once the template is complete, include it with the RFP and send it to the three or four suppliers that best fit the needs of the project and are still in the bidding process. Allow them one to two weeks (shorter is better) to respond with their initial bid.

Evaluating supplier responses is important and must be done with minimal bias injected into the decision. The Decision Matrix will help evaluate the supplier responses objectively. The SRM function will insist suppliers are selected on an objective basis. The Decision Matrix should help evaluate RFP responses for a specific project. Typical criteria include total cost, annual productivity, innovation, supplier capability/experience, etc. Table 11-2 is an example of a simple Decision Matrix.

Table 11-2 Simple Decision Matrix

Project # 1		Alternative Suppliers					
		Supplier A		Supplier B		Supplier C	
Criteria	Weight	Rating	Score	Rating	Score	Rating	Score
C1	2	1	2	3	6	2	4
C2	3	2	6	2	6	3	9
C3	5	1	5	2	10	3	15
Total Score			13		22		28

Each element should carry a "weighting" relative to other elements on the template. For example, "total cost" may be given a weight of 30% of the final score whereas "supplier training" may be given a weight of only 5%. The weighting recognizes the importance of each criterion – in this example supplier "C" achieved the highest score with 28 points.

Let's get real for a moment. Every member of the evaluation team will engage in the scoring process with prejudice. In fact, executive leadership may already have a preconceived opinion of who the supplier will be. The Decision Matrix should be used as a guideline to help the team evaluate suppliers as objectively as possible.

Unbiased scores will do two things: First, they may validate assumptions about the use of certain suppliers. Second, they may actually create a new/different level of awareness in those team members that are highly biased. A cross-functional evaluation team can mitigate some bias.

I have used a two-step approach to scoring. First, the Decision Matrix is circulated to each member of the evaluation team to enter a supplier score for each element on the template. Next, the selection team meets together to discuss and calibrate their scores. Someone needs to make sure the relative weight of each scorer is incorporated into the answer as well. For example, if selecting a vendor for a broad number of domains, a large domain should carry more weight than a small one.

➢ Discussion helps create a common understanding about the supplier and their cost estimates, time lines, resources, capabilities, etc. This is critical for final selection.

➢ For this to work, each team member must participate in every supplier interview – so keep the team size manageable.

Once the team review is complete, the suppliers can be ranked according to their scores. The top 2-3 suppliers should be invited to attend a joint Q&A session with the selection team and the other finalists. The objective of this session is to:

➢ Share the *same* information with each supplier to ensure a level playing field. Allow the suppliers to ask questions about aspects of the RFP that were not clear to them. This information is immediately available to the other supplier(s) in the meeting eliminating the need to send updates.

Note: As new information arises in subsequent vendor meetings it is fair to inform the other vendors you already met with about this new information.

➢ Openly compare how the suppliers react with each other and get a feel for how well they relate in general.

➢ Always keep at least two (three is better) suppliers in the mix EVEN if its obvious to your company that one vendor is in the lead. This is crucial to ensure competitive pricing – otherwise you lose your bargaining power.

The final step in the RFP process is to award the bid to the most worthy supplier. The team should make this selection soon after the joint meeting and inform the winning supplier. Once the winner has been notified and agrees to terms, those suppliers that did not win the bid should also be notified and thanked for their efforts. Further conversations with the supplier are needed to refine the elements to include in the SoW. Additional contract inclusions such as terms, SLA's, timelines, etc. should also discussed.

Note: There are a number of external consulting firms that specialize in helping clients work through the Request for Information (RFI) as well as the Request for Proposal (RFP) processes. They can be especially helpful if the in-house expertise in this area has not yet been fully developed. Their focused expertise can save the client time and money (hey, isn't that what outsourcing is all about anyway?). A very large initiative will especially benefit from the structured approach and extra boots on the ground.

If there is an existing MSA with the supplier, the next step is to draft a SoW. If an MSA does not exist, the focus of the next chapter is how to create one.

Chapter 12:

Master Services Agreement (MSA)

I must reiterate one more time that I am not an attorney and in no way wish to be mistaken for one. This chapter is informational providing a description of various business-related items to consider in the MSA. Always engage professional legal council to assure the best interests of your firm are represented when creating any contract with the supplier. Note: It is good practice to have the MSA drafted and signed well in advance as it is the primary legal framework. Requiring the supplier to sign an MSA prior to bidding on the RFP will expedite the bidding process and ensure basic terms are agreed to. The Statement of Work will contain project details that come out of the RFP.

In this chapter we will touch on common inclusions found in the typical MSA and leave it to you and your legal team to draft the legalese for elements you wish to include in the contract. These are noted in Table 12-1. There may be other topics added to your MSA that are germane to your organization and/or situation.

The objective of the MSA is to create an efficient framework for the supplier and client to operate within. The MSA describes the general Terms and Conditions that will govern the engagement between the client and supplier for all future work. A specific statement of work (SoW) will need to be created for each future project highlighting the unique aspects of that project. The efficiency comes from the fact that language contained in the MSA is automatically in effect for the SoW without having to replicate the language in the SoW. For example, the MSA with the supplier may contain language to define Intellectual Property, Hourly Rates, Invoicing, Confidentiality, etc. The MSA will usually state that the supplier is to provide service offerings in accordance with the terms and conditions of the MSA and one or more project agreements. This allows the client to focus on the specifics of each future project rather than replicating the general Terms and Conditions of the MSA. The MSA is the general agreement between the client and supplier.   The specifics for each project will be found in the SoW. We will explore the SoW in the next chapter.

Table 12-1 Master Services Agreement (MSA)

Strategic Concerns	Tactical Focus
Objectives of the MSA	- Defines broad reaching aspects of the contractual relationship between the client and supplier
Common Inclusions	- General "boiler plate" language - Services provided - Key personnel - Work approval and metrics - Financial terms - Confidentiality - Location - Changes to agreement - Data management and security - Intellectual Property (IP) - Statements of Work - Client responsibilities - Management and Control - Termination procedures

The Terms and Conditions in the MSA define the rights, protections and general business relationship between the client and the supplier. What follows is a practitioner's description for common inclusions found in the typical MSA.   Your legal staff should provide their opinion and legalese for the elements for your MSA.

"Boilerplate" is a term used to describe uniform language used in legal documents that has a definite, unvarying meaning. It is legal jargon commonly included in all contracts – not unique to IT.

➢ Recitals define who the parties to the contract are and general expectations for the agreement.

➢ Terms define many conditions contained in the MSA such as "addendum," "customer," "deliverable" etc. Terms are defined in the MSA and provide consistency for future agreements.

➢ Indemnification explains how each party may be protected by the agreement from certain legal actions.

➢ Assignment defines what happens if the supplier (or client) is acquired by another firm.

➢ Limits the amount supplier and/or client can be held liable for, except intentional misconduct. For example, supplier liability may be limited to the amount of fees paid by client per month.

➤ Warranties assure the service will meet certain specifications.

➤ Proprietary Rights explain the ownership rights of the client to all work product of the supplier performed under the contract.

➤ Force Majeure is temporary relief of responsibilities relieving the parties from liabilities for disasters beyond their control such as Acts of God, fire, war, terrorism, riots, etc. However, if these affect the supplier, they must show good-faith effort to enact any and all Disaster Recovery protocols and Business Continuity Plans they agreed to deliver under the agreement.

➤ Provision of Services will explain future project agreements added via a SoW. Project agreements explain how each SoW will be structured and defined for all future projects requested by the client.

➤ Performance definitions explain how the supplier's work product will be monitored and measured by the client.

  o Implementation and documentation responsibilities.

  o Defines language used for the work product (i.e. English, Spanish, etc).

  o The process for creating Service Level Agreements (SLA's) and Operating Level Agreements (OLA's) should be defined in a general way within the MSA along with potential penalties when they are missed. Specific metrics and parameters for SLA's and OLA's would be part of the SoW. We will delve more deeply into these performance measures and how to manage them in a future chapter.

  o Change control procedures are defined for the checkout and check-in of code and final "move to production" protocols.

Personnel related concerns such as availability, capability and language proficiency are defined. Demand Management (Chapter 6) when executed properly will help ensure the supplier can get the skilled resources on board in a timely manner. Demand Management will build on the partnership by involving the supplier in the process so they can plan better and potentially manage down costs.

The general roles of the relationship manager are identified for the supplier and the client. It is not unusual for the supplier to agree to hold key employees in certain positions to help create a stable environment for a specific project. Although general guidelines may be described in the MSA, specific project roles would be found in the SoW.

➤ A dispute resolution process is defined for instances where individual resources are incapable of performing their roles.

➤ Background screening and compliance to it is defined which forces the supplier to be more cautious when hiring. Even though access and accuracy of searchable data in third world countries may be suspect, I fully endorse background screening.

➤ Sub-contracting procedures should be defined in the event the supplier needs to supplement their resources. Liabilities, screening, etc. all need to be defined.

➤ The Visa process should be reviewed. The supplier needs to explain how they will bring non-U.S. citizens into the U.S. to work.

➤ Turnover of the supplier's resources can be a major problem especially after the client subject matter experts are gone. It is wise to put language in the Statement of Work targeting supplier turnover at a minimal level and that all expenses to train any and all replacements are covered by the supplier.

➤ Include verbiage giving you the right to a final review and approval of all resources assigned to the project. Vendor skills and language capabilities must be acceptable or the resource is to be replaced at supplier expense.

Work approval may be both subjective as well as metrics based. The MSA should define a general process so all parties know if the supplier's work is acceptable and delivered in accord with desired timelines. There should also be recourse to follow when they don't. Specific project deliverables and timelines will change from project to project and should be defined in the SoW rather than the MSA.

Financial Terms address a myriad of topics from Rate Cards to discounts and are typically found in the MSA. These terms will govern future SoW's with the supplier. The Rate Card contains the hourly rates for various skill and experience levels for resources in each of the supplier's geographic locations. For example:

➤ Java coders in India may be billed at a rate reflecting the local market while the same skill in Chicago may cost two to three times more. It is common for IT projects to require many different skill and experience levels. There can often be 10-15 different resources working across multiple geographic areas.

➤ Rate Card rates apply to future projects the supplier engages. The rates for the each skill levels and geographic location are multiplied by the planned number of resource hours. This forms the basis for the labor cost portion of a project SoW.

➢ There is often an annual rate increase built into the Rate Card to reflect escalating labor costs the supplier is experiencing. The rate increase is typically based on the rate being charged, not the actual wage being paid to the resource.

Everybody loves a discount! There are many ways to construct a discount program with the supplier. Some are easy to administer while others are quite complex. Some result in actual rate decreases, while others will generate a rebate to be paid to the client at some future date. All are subject to a negotiated agreement between supplier and client.

➢ One of the easiest discount programs to manage for both the supplier and the client is based on what is commonly called "Volume Slabs." Simply stated, a negotiated rebate amount is calculated for various levels of total payments made to the supplier during a stated timeframe – usually a year. For example, the following discount rebates might apply to the total amount of money spent with the supplier annually.

> $0 - $5MM of annual business will generate zero rebate
>
> $5MM - $10MM annually will receive a rebate of 5%
>
> $10MM - $15MM annually will receive a rebate of 7%
>
> $15MM - $20MM annually will receive a rebate of 10%
>
> $20MM - $25MM annually will receive a rebate of 15%

It is easy to manage because it is based on actual volume (dollars paid), which is easy for both the supplier and client to track. The rebate is paid on an annual anniversary date each year. This creates an incentive for the client to increase the size of the rebate by moving more business to the supplier.

One of the more complex discount programs was based on the client requesting a rebate amount paid "up front" in exchange for a promise to spend a fixed percent of their annual IT budget with the supplier. Although it sounded great and would have helped to solve a budget gap, it presented many issues. Here are a few:

➢ Defining precisely what the annual IT spending level will be is difficult. In addition to development costs does it include maintenance, administrative, support of the infrastructure, depreciation, etc? Does it include salaries paid to internal resources or just what is paid to third party suppliers?

➢ How will the supplier validate the client's promise? How will they know what was actually spent? Is the client willing to open their books to the supplier – this would include showing what was paid to other IT providers. Not very practical.

➢ Assuming the rebate is paid prior to the actual IT spending occurring, what happens when business conditions change and IT is forced to reduce the amount it spends?

➢ What happens when a change in business strategy requires that more money be spent with a different supplier?

➢ I do not endorse this type of discount program as it has far too many opportunities to generate doubt and destroy trust between the parties. Believe me, it is much easier to use the "volume slab" method.

Payment terms pertain to how the invoice will be generated and how soon the client must pay the invoice from the time of receipt. This is usually a straightforward section in the MSA and, but like everything else, is open to negotiation.

➢ Generally, suppliers invoice the client on a monthly basis. The monthly invoice should include a line item for each of the active SoW's the supplier is engaged in with the client.

➢ The client will be given a certain amount of time (i.e. 60days) to pay the invoice from the time it is received.

➢ The MSA may also include a workflow description of the invoicing and payment processes.

Confidentiality agreements are core to all contracts between clients and suppliers. The intent of course is to assure that information and data central to the client's business is not shared with other companies the supplier may work with. In some cases, the confidentiality verbiage also protects the supplier from the client sharing information about pricing and work elements the supplier does not want shared with others. The MSA should contain descriptions of exactly what is meant by confidential information and for how long. Penalties for any breach of confidentiality should be defined in the MSA.

Geographic location drives the cost of labor that makes outsourcing offshore a financially attractive option. As long as SLA's, KPI's and OLA's are met the client really shouldn't care where the work is done. On the other hand, all geographies are NOT alike and other factors may overshadow any labor arbitrage objectives. Geographic location is included in the MSA to allow clients the opportunity to agree (or disagree) with the location strategy of the supplier. Here are a few factors for the client to consider:

➢ Geopolitical stability of the region. In a country (or a broader geographic region) where the government is continually facing crisis, the risk of the work not being done as intended increases. One needs to question the wisdom of investing time, money and effort to build an outsourcing partnership in an unstable environment?

➤ Personal safety is another factor to consider. Remember, at some point the client really should visit the offshore location to bolster the relationship. Even though the governments may be stable, there are some countries and cities that present a significant danger for travelers. For example, Brazil is a stable country with a booming economy; however, visiting certain areas can be an unsettling experience.

➤ It is important to review the education infrastructure of the country. Are they capable of producing enough high-quality grads to assure current and future needs can be supported? When demand outstrips supply as it often does in India, wage inflation and high turnover will soon follow.

At some point, the existing MSA and/or subsequent SoW's will need to change. New services may also be added by mutual consent of the client and supplier. To the extent possible, creating language for how changes will be managed and how new services would be "transitioned in" will help the process go more smoothly. Some changes to the MSA can be triggered by language within the MSA itself. For example automatic annual rate card increases can be baked into the contract. These are usually based on an agreed to government issued economic indicator such as the Consumer Price Index (CPI) or a similar statistic. Modifying the MSA by mutual agreement of both parties is usually spelled out in the contract. This includes new services; a new project, or changes to the language in the indemnification clause.

The SoW for application support can span multiple years and should contain language needed to modify the initial scope of work. This will be needed when adding new applications to the scope or eliminating others. New applications tend to have high incident rates shortly after rollout. The incident rate slows over a 6-9 month horizon as bugs are fixed and users learn the application. It is wise to wait for the application to settle down before determining how many resources are needed to support it. I advise clients to manage the first 6-9 months of a new application as a separate Time & Materials contract until the application reaches a more stable state. Once stabilized, the client and supplier have a better idea of the resources needed to support the application.

The amount of work required to support "in-scope" applications may change as well. An application may be eliminated or improved to the point where the incident rate diminishes. Both of these events will reduce the resource needs. On the other hand, applications may have new functionality added or are rolled out to a broader user base. These actions will increase the number of incidents and the subsequent need for more support resources to address them.

The contract should define how to modify the monthly invoice to address a resource being added ARC (Additional Resource Change) or reduced as an RRC (Reduced Resource Change). The supplier will be good about requesting ARCs. You should ask to see all pertinent data that supports their claim. In fact, it is a good idea to review resource deployment on a monthly basis to understand how many support hours are being charged to various application clusters. This information will help the client understand the Total Cost of Ownership (TCO) of their major applications and consequently how much they could save if certain applications were improved or retired. It is critical to understand the assumptions and capability to support the original scope.

➢ Reviewing resource deployments on a year-over-year basis is a good practice to share with the Business Partners to show where support dollars are being spent. As a component of Total Cost of Ownership, this will help develop the strategic plan for major application groups.

Language to assure that proper protocols are in place for data management and security are also found in the MSA, as they will be required for every future project. Included are descriptions of:

➢ Data ownership – generally the client is named as the data owner but either party can perform data management roles.

➢ The client defines data retention policies for the supplier to follow.

➢ Application security includes processes to assure:

  o Application code cannot be changed without approval

  o Application data is protected from theft or destruction

  o Audits at supplier locations are routinely performed to assure compliance to security protocols.

The supplier must be required to actively participate in routine Disaster Recovery (DR) exercises in various different ways. SoW's for support will contain very specific details about the mutual DR roles the client and supplier will perform. These will usually consist of testing and confirmation that applications have come up on the back-up server and are producing valid results.

The supplier should also be asked to provide a description of the DR process on their side of the communication link. Indian firms typically will have data centers in India with a communication hub somewhere in the U.S. They need to show how their DR protocols work from the U.S. hub to their servers. Offshore resources will be managing a large amount of work remotely. When the communication link breaks the work stops!

Intellectual Property (IP) is the unique core business knowledge the firm uses to drive their business results. The firm or supplier may suffer significant financial loss if their IP falls into unscrupulous hands. The MSA language pertaining to IP should be sure to address any or all of the following:

➢ Patents, trade secrets, know-how and proprietary information.

➢ All copyrights, copyright registrations and applications across the world.

➢ All industrial designs, registrations and applications across the world.

➢ All rights to World Wide Web addresses, domain names and registrations.

➢ All trade names, logos, common law trademarks, service marks, trademark and service mark registrations and applications throughout the world.

Depending on the business sector, there may be industry specific IP requirements. Example: in the retail sector Credit Card companies require that rules of the Payment Card Industry (PCI) are followed to assure credit card transactions, consumer privacy and financial exposure are protected.

IP rights for "vendor software" (applications or tools purchased by the client for use by the client such as mainframe monitoring) should not be ignored. It is not unusual for clients to have purchased hundreds of vendor software products over the years. Recent purchase agreements are more likely to include language allowing a third party to use the product. However, many purchase agreements specifically bar third party resources from using the software. Unless the use by a third party is specified and approved in the contract with the vendor, the client may need to seek approval from the vendor for the supplier to use the application – most vendors will be happy to grant this approval for a fee. Failure to secure a "Right to Use" (RTU) approval from the software vendor may result in damages if they decide to pursue action.

> Searching through hundreds of vendor contracts can be daunting. The addresses change and many are swallowed up by or merged with other vendors. The outsourcing supplier wants to avoid any legal hassles with the software vendors and may request that an indemnification clause regarding RTU be added to the MSA to hold the supplier harmless.

The language should indicate that the client has exclusive ownership of all documentation and code created by the supplier during the engagement. When this language is missing, nasty consequences can follow. I had a client that did not bother to specify the ownership of code that one of their suppliers developed for them during the engagement. The supplier actually charged the client for the ongoing use of the code – and their case held up in court! Since you will be paying for it, make sure you own it!

Statements of Work (SoW), as mentioned earlier, are specific projects that fall under the framework of the MSA. Unless specific exceptions are noted, all elements of the MSA will apply to any future SoW's performed by the supplier. We will get into the details of what should be included in the SoW in the next chapter. The MSA defines the structure within which SoW's operate.

> Note: In SoW's pertaining to application support, the term "nonrecurring initiative" (NRI) may also be used to describe a subproject that may be the work product of the Support Team but requires incremental funding. These are generally small, short-term projects. General aspects of the NRI such as installation and testing are also included in the MSA.

The MSA also defines the roles and responsibilities to be performed by the client. Examples of obligations the client may agree to include: providing computer workstations, office space, parking, security, etc. The client may also be asked to provide a safe and secure environment for the supplier's resources to work in.

A variety of management and control activities may also be included in the MSA. Governance is a client role. It is the culmination of activities that assure "value leakage" is held to a minimum. Good governance makes sure the supplier delivers the value they agreed to deliver in the contract.

General terms concerning performance reporting will be explained in the MSA. Typical reports will indicate where resources are deployed by application group (i.e. 15% focus on HR Systems); the number of problems resolved and more importantly, the number that are not yet resolved; chronic incident points and suggestions for resolution; and many more. The specifics should be noted in the SoW for each project.

Requirements for meetings are also defined. Meetings may cover multiple topics and occur at various levels across the firms. Examples include:

➢ Weekly meetings at the Process Team level to discuss issues that arose during the week and how they were addressed; service failures and how to prevent them from recurring; staffing training and KT progress.

➢ Monthly meetings with the Working Team include reviews of how well the project is tracking against SLA results; project updates on milestone attainment; status update on minor enhancements and more. Knowledge and record retention are topics to be addressed periodically throughout the year.

➢ Quarterly meetings with the Executive Steering Team may be used to discuss market trends; innovation ideas; overall performance and business partner perceptions.

Change Control describes the process for how new code will be created, tested, and moved into the production environment. This will contain a number of necessary security protocols.

Nothing lasts forever - business dynamics change, firms merge, relationships sour. Termination procedures should always be part of the MSA so the process can be agreed to well in advance of ever having to execute it. It is much easier to negotiate these when the relationship is on a high note. Trying to negotiate a termination agreement once the relationship has deteriorated is close to impossible. Termination language defines how to undo the contract or portions of it. Items such as:

➢ The contract should contain a definition of what is meant by "termination for cause".

➢ Knowledge Transition (KT) from incumbent supplier to new provider.

➢ Notification period – how far ahead should the supplier be notified when a specific resource will no longer be needed by the client.

➢ Penalties for early termination.

   o The supplier, as well as the client, has invested a lot of time and money in the relationship so breaking the contract early in its term may not allow the supplier to recoup all of their investment. It is quite common that the client is required to pay a penalty to the supplier for breaking the contract early. The penalty and its size are always open for negotiation.

- o "Termination for cause" implies the supplier has demonstrated they have not and will not provide the skills and experience needed to execute their commitments. Evidence of this may be seen in SLA deterioration and/or an increase in complaints from business partners. When there is cause for dismissal, the supplier should not be granted an early termination penalty.

Defining the KT process is critical. Replacing a supplier is a decision that should not be taken lightly as it is both costly and risky. The existing supplier may be in place for a number of years before the decision is made to replace them. Over time the supplier becomes the subject matter expert. Success therefore, requires that the existing supplier transfer their knowledge to the new supplier or back to client resources. Expect problems from suppliers in the shallow end of the gene pool. Few are happy to help another take their business away. Top tier firms on the other hand, realize that someday they may have an opportunity to win your business back and will not impede the KT process. You win or lose by the supplier you chose!!

➢ One of my clients had been using Supplier X for a number of years for both application development and support. Their individual resources from Supplier X were high functioning and well thought of. The problem was with their cost structure and the fact they were primarily used in a staff augmentation mode. To make matters worse, they did not document changes made to the code or environment and their process management capabilities were lacking. They were not happy when they were given notice that they would be replaced. Initially they refused to allow the new supplier into their India-based facilities for KT training. After some arm-twisting by our CIO, they reluctantly agreed to give the new supplier access to their facilities.

The next chapter will focus on the Statement of Work and how it relates to the MSA.

Chapter 13:

Statements of Work (SoW)

In the prior chapter, the Master Services Agreement (MSA) was discussed with heavy reference to the Statement of Work (SoW). In this chapter we will follow the thoughts in Table 13-1, which provide a detailed explanation of the objectives and common inclusions in the SoW. The MSA provides a general framework for the contractual relationship between the client and supplier. It is an agreement on how to operate but doesn't address specific projects – the SoW is used for that. If there were no MSA, every SoW would need to replicate all elements described in Chapter 12.

Table 13-1 Developing the Statement of Work (SoW)

Strategic Concerns	Tactical Focus
Objectives of the SoW	- Once the MSA is in place, the Statement of Work defines the expected outcomes for each subsequent project with that supplier.

Common Inclusions	- Define project scope and desired outcomes/capabilities
	  - Identify major milestones
	  - Work with the supplier to establish timelines
	  - Define activities to be performed by the supplier
	  - Agree on metrics to measure deliverables
	  - Name technologies, platforms and specific applications
	  - How the SoW will be added to the main contract
	  - Manage as a non-recurring initiative or ongoing work
	
	- Alignment of activities
	- Geographic locations
	- Named Resources
	- Knowledge Transfer process
	- Costs
	- Equipment

The objective of the SoW in a Managed Services relationship is to provide definition of all inputs and outputs for each subsequent project the client and supplier engage in. It is important to be as detailed as possible with the facts in the SoW. Project timelines and milestones are core components in the SoW as are cost, quality expectations, metrics and business outcomes. Be sure to work closely with the supplier to assure they understand what is required and agree on the outcomes, inputs and quality levels for the project.

Managed Services SoW's generally fall into four common categories. There are definite similarities across these categories; however, specific details are still required for each SoW. The four types of SoW are:

➢ Application Support – the supplier agrees to provide break-fix and other defined support activities. Support SoW's are designed to span a number of years. There are different SLA's, metrics, etc. than other SoW's and usually are associated with a specific defined scope of applications.

➢ Infrastructure Support – These address the need to manage the physical hardware, telephony, data centers, etc. of the enterprise. These are also multiyear agreements.

➢ Help Desk – in this type of SoW the supplier agrees to manage all aspects of the client's Help Desk including the Call Management Center. This too is usually a multiyear contract.

➢ Application Development – this engages the supplier in the creation of new code providing enhanced capabilities for the client. The span of this SoW includes: logical/physical design, coding, testing, and installation. The time frame for a development SoW is nonspecific and determined by the size of each project. The SoW ends when the project is completed.

Even though the Terms and Conditions have already been drafted in the MSA, it takes time to get SoW's approved by legal. This is because each SoW is a unique agreement due to various project details. One suggestion to help speed the approval process is to work with legal and create "SoW Templates" for the IT project lead to fill out. The template prompts the project manager to provide only those variables that make the SoW unique. All necessary structural components are part of the template. It becomes a repeatable process in which the client focuses on important project detail and not general contract language. Once the use of a SoW template is adopted, a cursory legal review is all that is required.

The SoW describes the specific scope and outcomes anticipated with the project. These would include defined milestones; timelines; activities performed by supplier and by client; metrics; technology platforms; language to enjoin the SoW to the MSA; Nonrecurring Initiatives and project management structures.

In a Managed Services relationship, SoW's must contain a description of the outcomes the project will deliver. Here is a minimal example describing an "outcome" that might be found in a development SoW:

➤ XYZ Corp wants to engage Supplier X to develop an inventory management application. The application must provide a perpetual inventory plan by day for 3 weeks into the future with an updated daily view of available inventory on hand for each item; actual and expected replenishments for each item by day; forecast and actual demand by day; and a calculation of "days of supply" (open divided by forecast). The scope will include the 250 items in the Dog Food Business managed across 4 distribution centers and 2 manufacturing facilities.

➤ Milestone dates for requirements definition; code design; construction; testing; transition and implementation should be established along with target dates for project completion.

The SoW for an application support project will define specific work and outcomes including:

➤ Activities to be performed by the supplier such as: application break-fix, minor enhancements, application maintenance, etc.

➤ A description of the primary Service Level Agreements (SLA's) and other metrics used to monitor delivery. Specific standards for these metrics will be established once enough data has been collected to describe process capability. This will be discussed more in the next chapter.

➤ Supplier's role in Disaster Recovery testing, Business Continuity and Sarbanes-Oxley (SOX) compliance.

➤ Technologies, platforms and specific applications that will be part of the project touch points.

➤ Scheduled meetings and required reports between the client and supplier to discuss the delivery of productivity promises made by the supplier, additional cost reduction opportunities, innovation ideas, etc.

➤ The actual definition of outcomes would be much more exhaustive than what is noted above. The description should be rich enough to give the supplier a fundamental understanding of the work product to be delivered.

Note: Managed Services SoW's differ quite a bit from those used with staff augmentation model. Under staff augmentation the client requests resources with specific skill sets and directly manage them to deliver the work product. The client is responsible for the application – development, delivery and ongoing support.

In order to track progress, milestones are needed. These should be defined at a high level, and jointly created between the client and supplier. It is not uncommon for suppliers to be penalized or rewarded for their ability to miss or achieve major milestones. Examples for a development project include:

➢ Requirements gathered; initial design approved by May 1

➢ Construction and unit testing completed by Aug 1

➢ Integration and user testing completed by Oct 1

➢ Installation and rollout to begin by Nov 1

➢ Transition to support to occur by Jan 1

The supplier leads the effort to create a detailed timeline for all of the sub-tasks required to deliver the major milestones on time. Client project leads will also participate to assure things are moving in the right direction. The detailed timeline should be reflected in the SoW. Identifying technology platforms and specific applications integral to the SoW will eliminate confusion and help the supplier identify the specific skill sets required for the project.

There are a number of metrics available to help monitor the flow and timing of work products associated with the SoW. Tracking actual days used vs days planned is a good way to keep on top of things. Another metric could be the defect rate (QA test failures) per module or lines of code.

New SoW's will be incorporated into the existing contract structure based on language in the MSA. However, there may be a problem accounting for payments upon completion work products and/or services delivered. Although there is only one MSA with the supplier it is often the case that many IT teams may work with that supplier simultaneously resulting in multiple SoW's. In IT shops with one budget, all SoW charges hit that budget. However, when there are multiple budgets, it is important that the invoiced charges from the supplier land in the correct funding source, or cost center. The SoW must be clear about which cost center the invoice will charge. Cost Center budgeting and actual results vs the budget should be monitored closely to assure that the money flows accurately through the organization.

➢ The SoW should be clear about payment terms and whether the project is to be fixed fee, pay on delivery of milestones or time and materials. Refer to Chapter 4.

Earlier we introduced the term "nonrecurring initiative" (NRI). NRI's refer to work done outside the terms of an existing SoW. For example, let's say that Supplier X manages the support desk and performs break-fix, and other recurring activities. However, a major coding defect has been discovered that if corrected, would greatly improve the application. The size of the project to correct the defect is greater than what can be absorbed by the Support Team. This non-recurring event will require its own SoW followed by an invoice once the work is delivered.

For large initiatives, it is a good idea to include discussion of the management structure in the SoW. This will include specific leadership positions provided by the supplier as well as the client. To maintain continuity, there might also be limits to changing these positions during the term of the project.

Most IT shops work on multiple projects at the same time. Because of this, it is important to make sure that these projects are aligned across all major work groups in the organization. Too often, project teams act as if they were in a silo ignoring other teams. Unfortunately, they will need to interact with other teams at some point for project to be successful. As the milestones and specific deliverables are defined for a SoW, it is critical to review these with all stakeholders (both suppliers and internal IT teams) that will need to participate in some way on the project. This may result in the need to modify and adjust work plans and goals.

Infrastructure teams need to be made aware of the project to assure that new environments, servers, etc. can be ordered and installed in time to meet project timelines.

Database analysts will be called upon to provide work on the project. They need to understand magnitude of their role and timing for when the work needs to be completed. Like other teams with limited resources, their work needs to be scheduled across the many other projects in play.

➢ Be sure to include those applications that the project will need to integrate with. A subject matter expert (SME) familiar with that application will need to do some work to help build and test the interface. Their work requirements need to be aligned with their other responsibilities and commitments.

➢ The timing for Knowledge Transfer (KT) and user training must be identified so resources can be scheduled to complete these tasks.

One of the more difficult aspects of project management is keeping all teams abreast of changes. Unplanned surprises cost money!! Example: If the DBA's have set aside time to work on "Project A" from Oct 10-Oct 30 their resources will be idle if the project slips by a couple of weeks without prior notification. Conversely, if the project is ahead of schedule, the DBA's may not be able to drop what they are doing to complete their work in turn may idle other resources on the project. I think you get the drift – good planning, re-planning as required and excellent communication of the plan will reduce waste and improve delivery results.

Geographic location of project resources should also be noted in the SoW. With good SLA's in place, the geographic location shouldn't matter. However, you need to understand <u>how</u> the supplier will manage and coordinate work efforts across multiple geographies on a "follow the sun" basis. Managing resources between the U.S. and India is a challenge and exhausting with daily late night conference calls with the offshore team. When a third geography is introduced, the issues are magnified. It is easy to conceptualize but difficult to implement the coordination of resources located in multiple time zones across the globe. Further, if an "agile" development will be used, it creates even more challenges managing project resources based in remote locations.

➢ My advice is to be open to geographic suggestions but press the supplier to explain how they will manage those resources to deliver the SLA's and business objectives important to the specific project. The supplier can arrange a reference call with other clients that are utilizing a multi-geography approach.

Knowing the skills and background of resources the supplier intends to use on the project is good. Again, with solid SLA's in place, it shouldn't matter. However, there are certain things to make sure you have outlined in the SoW.

➢ The client and supplier need to agree on the technical and managerial skills required for project. Understanding and agreeing on those skills will provide a good insight into how closely aligned both sides are to the needs of the project. If the client thinks the SoW is a huge effort, but the supplier believes it to be a chip shot, the two sides are not aligned very well.

➢ Language competency requirements may be different for each SoW. For example, in application support, the user will frequently talk directly with the support resources and need to understand them. On the other hand, for a development project, many of the offshore resources only talk with other supplier resources. Language competency is not as important when there is limited interaction with client resources.

➢ A common indicator of capability is "years of service." It's OK to have a couple of junior resources on the project; however, the supplier needs to commit to a certain level of experience across the team. Something like "on average, the team will have 7 years of experience." There should also be a defined minimum level of experience too "no member of the team will have less than two years of experience." Caution: Although it's easy to measure years, it is hard to measure the depth of experience and competence gained over those years.

Onshore to offshore ratios will determine the cost of the project. If the supplier offers an acceptable fixed bid cost estimate, these ratios become less important. High offshore ratios result in better cost, but may create incremental hurdles for the client to deal with because of the distance and time zone offsets.

Key resources for the project are often identified along with the supplier's commitment to keep them on the project. Turnover is inevitable so any opportunities to limit it should be taken.

Solid Knowledge Transfer (KT) is critical for the outsourcer to have the information needed to successfully manage their assigned project tasks. KT is most important for application support projects. This type of project is intended to reduce overall operating expense and this is often accomplished by eliminating incumbent staff. Incumbent resources are typically the "subject matter experts" for the applications being outsourced. Without a solid KT from incumbent to supplier, vital knowledge will be lost in the transition putting support capabilities at risk.

I have found a great deal of application knowledge is non-recorded "tacit knowledge." Tacit knowledge is acquired via experience. The individual learns through doing and rarely records the solution in any usable form. He or she simply reacts to an incident. "Explicit knowledge" on the other hand is often recorded in a searchable manner giving the practitioner quick access to solutions of the past.

As the supplier takes over the support role, they become the subject matter experts. It is important that they document all fixes and changes to the application in order to build their knowledge base. Of course reliance on tacit knowledge will still exist but eventually be reduced as more and more is recorded. Shifting the process to rely more on explicit knowledge will help minimize the amount of tacit knowledge needing to be transferred to future generations.

Equipment needed to execute the project should also be included (or specifically excluded) in the SoW verbiage. In today's world, the practice of locking down the image for company owned assets such as workstations is commonplace. The client usually provides workstations for resources working at their site. This is done to minimize risks associated with replicating the client's image on supplier owned workstations. Other items include desks; private office if needed by project leads; cell phones; use of client facilities such as the cafeteria and exercise facilities; and other amenities.

Multi-year support contracts must have verbiage in the SoW to address ongoing productivity and innovation and how to measure them. Be sure to include tangible performance targets to measure productivity such as an annual cost reduction percentage (5% is typical); the number of incidents by application cluster; total number of resources used; the number of minor enhancements delivered; and the ratio of onshore to offshore resources.

Innovation is more difficult to measure objectively. Although suppliers must be innovative to achieve their objectives, most clients think of innovation as major "game changing" ideas. The world-class suppliers I have worked with have a wealth of experience with the technologies and business processes in play within your market. They can be a fertile source of innovative ideas that, in time, may become that "big" idea. Conducting quarterly reviews with the supplier is a good way to establish a dialogue with upper management to generate ideas that will help to boost tomorrow's bottom line results.

> As an example, three years before it became commonly accepted, one of the suppliers I had been working with described a technology for retail that would send an alert message to the customer's SMART phone. The message would tell the customer about special offers tailored to their buying habits as they approached the store. This new type of marketing technology was able to increase sales and drive bottom line results.

Cost estimates and the assumptions that shape them are also described in the SoW. There are two primary cost drivers involved. The first is the infrastructure cost for servers, workstations, telephony, etc. The other is labor to design, code, test and install the project that is highly sensitive to the location of the resources. Another aspect of the cost estimate for a support SoW is the term of the contract. Even though most support contracts have clauses allowing early termination, some corporations look at the annual cost times the number of years to determine the financial amount "at risk" for the firm.

As we have noted before, it is important for the infrastructure team to be made aware of the hardware and connectivity requirements for the project. One of the outputs of the infrastructure team is a cost estimate for the hardware configuration. It is assumed that the application design fits with the firm's architectural standards.

Many factors including the number of resources needed; skill levels; and location drive labor cost. Labor is usually the single largest cost element of IT projects. Chapter 5 went into great detail explaining the cost advantages that outsourcing may deliver. I won't bore the reader replicating that chapter here but will point out that project costs should include training, KT, travel, client staff time, etc. Suffice it to say that an effective outsourcing strategy will have an important impact on costs.

The inconvenient truth about cost reduction is that you can only be a hero once! Let's assume that an IT organization has an annual operating budget of $100MM and a major outsourcing initiative reduces costs by 15% or $15MM dollars. This is a great improvement. However, the budget also gets reduced by 15% and the new budget becomes $85MM. Next year, the IT management team is in the same budgetary boat they were in before the cost reduction efforts were enacted. Worse yet, expectations have been planted that IT can deliver 15% reductions ad infinitum. This is a ridiculous assumption because there are very few additional levers to help meet ongoing cost reduction goals.

The *Inside* of *Outsourcing*

Chapter 14:

Performance Metrics

When working with third party partners, it is obvious that tools are needed to measure and track supplier performance. Not so obvious is exactly what needs to be measured; where the data comes from; and how to present the data in a way to identify service issues when they occur. Service Level Agreements (SLA's), Operating Level Agreements (OLA's) and Key Performance Indicators (KPI's) are the 3 primary types of metrics used to measure supplier performance and service quality.

Table 14-1 Service Level and Operating Level Agreements

Strategic Concerns	Tactical Focus
Objectives	- Metrics help to monitor and evaluate supplier performance
Service Level Agreements (SLA's)	- Typical SLA's used to measure application support - Definitions needed for up-time; user satisfaction; minor enhancements; etc. - Create access to the ticket management system; project documents; etc. - Identify data sources - Define the priority and severity levels for each application - Reporting - Default penalties and earn-back opportunities - Adding/deleting SLA's - SLA's used to measure development projects
Operating Level Agreements (OLA's)	- OLA's ensure activities of multiple vendors are aligned to the SLA that most benefits the client
Key Performance Indicators (KPI's)	- Understand how KPI's differ from SLA's - Typical KPI's

The objective of performance measurement is to provide analysis, based on data, which describes how the supplier is performing. The use of observed data, rather than opinion, eliminates emotion from the discussion and helps the client and supplier reach common ground on critical failure points that require improvement. The analysis will also provide insight on how best to prioritize the implementation of process improvements. This is similar to the Total Quality Management (TQM) approach used to manage and improve manufacturing processes for years. In fact, many of the statistical measures and terminology will be familiar to those who have worked with and understand TQM.

Develop Service Level Agreements (SLA's) that are based on solid data and accurately define the services and quality expectations the supplier is to deliver. DO NOT PUSH THIS OFF AS SOMETHING TO DO AFTER THE CONTRACT IS SIGNED!! In a fixed-price contract, as long as SLA's are met, how the supplier staffs the work is unimportant. Be sure to understand assumptions for the mix of onshore to offshore and how this will change over the course of the contract. SLA's are the backbone of the Managed Services Model. Lead-time and frequency that SLA's can be changed should be spelled out in the contract to allow the supplier time to adapt to meet goals. Properly written SLA's will benefit suppliers when met, and penalize them when missed.

A common mistake is to try and measure everything. First, the resources required to collect and report the data are cost prohibitive. Secondly, a few key metrics is all that is needed to tell the story. The trick is to select metrics that are not easily "gamed." Also, one needs to be mindful to not create more problems because of the metrics used. The report card drives behavior. Here is a true story:

➢ The Customer Service team for a large food company was working hard to improve their order fulfillment capability. They established a very high Customer Service Level (CSL) target to track how they were doing. They measured "what was shipped" vs "what was ordered." The Sales Team dictated that a 99% CSL target be established. The Customer Service team found that one way to achieve that goal was to increase inventory. Unfortunately, a higher inventory level not only creates a higher carrying cost, but in the food industry, a greater risk of product spoiling before it can be sold. Trying to reach that high CSL measure actually resulted in negative impacts to the business results. To address rising inventory levels, two additional metrics were added. One was Days of Supply (DOS) - a simple ratio of on-hand inventory divided by the sales forecast; the other was Percent of Stock Sold at Discount (PSSD). The Customer Service team was required to achieve all three measures and therefore couldn't "game" the system. They had to find a better way to manage the process. They were actually quite creative and implemented the use smaller reorder quantities more frequently – instead of receiving one delivery every other week (covering two weeks of demand), they ordered smaller quantities to cover four to five days of demand and did this two to three times during that same two week period. The lofty Customer Service Level was achieved! Additionally, the customer actually received fresher product and when inventory levels dropped amount of distressed sales was virtually zero.

Create metrics and weight them in order to drive desired behavior by the supplier. We will discuss those used more frequently within an IT services environment.

Common metrics used to measure application support include:

➢ Uptime is a frequently used SLA that is simply shows the percent of time the application is available to the user. This is a good measure when executed properly. The essence of the metric is to determine how many hours the application is operational divided by the scheduled hours of use. For example, if the users need the system to be operational from 8am to 8pm, that would be twelve scheduled hours of use. If for some reason, the application were down for two hours during that time, the resulting measure would show 83.3% up time. If the measure is based on a 24-hour day, and was down for two hours, the resulting measure would be 91.7%.

➢ It is important to define the scheduled hours properly. This gets more difficult as the need for a broader window of availability gets wider. Most applications require some sort of routine maintenance that makes it unavailable for some period of time. Companies with globally deployed applications need to carefully find a maintenance window that minimizes disruption to the users. Given the fact that routine maintenance is required, some firms chose to deduct it from the scheduled hours; others will consider it a hit to the uptime metric and look for ways to reduce it's impact.

➢ Data for this metric primarily come from the Call Management Center (CMC) database. The CMC database will capture the time the application goes down, and when it becomes operational again.

➢ A run chart displaying uptime results by day is a good way to present the data. Once enough data points have been collected, it is possible to determine the mean, as well as variance about the mean. With that information, the TQM concept of Upper Control Limits (UCL) and Lower Control Limits (LCL) can be calculated. All subsequent measures can then be plotted to see how they land within the statistical control limits.

➢ It is important to understand that SLA's should be based on process capability rather than performance desires. If the data shows the application is only capable of achieving 90-93% uptime, it is a mistake to assign 99% as the target value for the SLA – since the underlying process simply can't deliver it! The process will not improve on its own; time and money are needed to change it and improve results. It is fine to establish 99% as a long-term goal for the supplier. The supplier can then take a close look at the underlying events and provide an informed estimate for what it will cost to achieve 99%. Assigning unrealistic standards does not make the process better. Most suppliers are savvy enough to not get caught in these traps. However, there are some that are hungry enough to agree to anything – be wary!

User Satisfaction is a metric based on "feelings" rather than fact. Users are asked to complete a survey every few months regarding their experience with the supplier. Questions address areas such as "responsiveness"; "professionalism"; "willingness to help the user"; and other touchy-feely types of questions. It is important to design survey questions carefully so as not to bias the results. Note: If you don't plan to take action in response to this survey don't do it all. Why set up false expectations for the user?

There are a number of SLA metrics designed to measure the time it takes for the supplier to provide certain services.

➢ Time to Acknowledge – is the elapsed time between when the service request (ticket) was posted with the Help Desk and when the supplier responds acknowledging they received and are addressing the incident. Typical standard times for this range from 15-30 minutes.

➢ Time to Resolve is the elapsed time between when the ticket was opened and when the incident is resolved and service is restored. The standard time for this metric is variable and based on the severity of the incident and criticality of the application to enterprise business results. For a highly critical application such as Order Entry, standard time may be four hours. For less important applications, the standard time may be up to one or two days.

Table 14-2 is a matrix of priority levels for applications across three tiers: Gold, Silver and Bronze. To the user, every incident they experience is very important and their initial response will be that every application should be top tier (Gold) and every incident should be P1 (critical). Even though the SLA metrics can accommodate this scenario, the cost would be prohibitive. Therefore, consensus (and some common sense) is needed when IT and the Business work to assign the appropriate tier level (Gold, Silver or Bronze) to each application. "Mission Critical" applications such as Shipping, Cash Registers, Order Entry, etc. are typically defined as Tier 1 or "Gold." The key differentiator for Mission Critical is that when these applications are down, the enterprise is losing revenue and cash flow is impacted. Tier 2 or "Silver" applications are considered "Business Critical" and an extended service outage will start to negatively impact business results. Silver applications include Payroll, HR Benefits, Accounting, Scheduling, etc. An application defined as "Bronze" is important to the enterprise but will not result in a cash flow or business management issue if the incident is not resolved immediately.

Table 14-2 Service Level Management

Performance Expectations	Time Frame	Type of Metric			Gold: Mission Critical			Silver: Business Critical			Bronze: Important		
		SLA	KPI	New	SLA	Expected	Minimum	SLA	Expected	Minimum	SLA	Expected	Minimum
Incident Management					30% of Application: Gold			60% of Applications: Silver			10% of Applications: Bronze		
Incident Ticket Time To Resolve													
P1 - (Critical)	Monthly	X			4 hrs	99.0%	96.6%	8 hrs	95.9%	92.5%	12 hrs	88.8%	61.8%
P2 - (High)	Monthly	X			8 hrs	93.8%	85.1%	21 hrs	96.2%	93.1%	42 hrs	90.1%	86.2%
P3 - (Medium)	Monthly		X		16 hrs	86.3%	72.8%	52.2 hrs	96.5%	96.0%	104 hrs	97.4%	96.6%
P4 - (Low)	Monthly		X		2 weeks	99.2%	99.2%	3 weeks	99.1%	98.5%	4 weeks	97.8%	94.8%
Customer Satisfaction	Monthly		X		Average	85.0%	80.0%	Average	85.0%	80.0%	Average	85.0%	80.0%
Problem Management													
Submit Initial Analysis													
P1 - (Critical)	Monthly	X			5 days	95.0%	90.0%	5 days	95.0%	90.0%	5 days	95.0%	90.0%
P2 - (High)	Monthly			X	10 days	95.0%	80.0%	10 days	95.0%	80.0%	10 days	95.0%	80.0%
P3 - (Medium) - Chronic	Monthly			X	20 days	95.0%	80.0%	20 days	95.0%	80.0%	20 days	95.0%	80.0%
Submit Final Analysis - Resolve													
P1 - (Critical)	Monthly	X			10 days	95.0%	90.0%	10 days	95.0%	90.0%	10 days	95.0%	90.0%
P2 - (High)	Monthly		X		20 days	95.0%	80.0%	20 days	95.0%	80.0%	20 days	95.0%	80.0%
P3 - (Medium) - Chronic	Monthly		X		40 days	95.0%	80.0%	40 days	95.0%	80.0%	40 days	95.0%	80.0%
Change Management													
Fixed right first time	Monthly		X		Result	95.0%	90.0%	Result	95.0%	90.0%	Result	95.0%	90.0%
Changes implemented on time	Monthly			X	On Plan	99.0%	95.0%	On Plan	99.0%	95.0%	On Plan	99.0%	95.0%

Not every incident within a tier will be critical. Priority levels (also called severity levels) need to be defined for the various types of incidents an application may experience. Some may be defined as P1 (critical) – example: "the shipping system is down and product can't ship to customers." Other incidents for the same application may have a lesser severity such as P3 (medium) – example: "a change to the text of the invoice header is needed." From Table 14-2 we find that 30% of all applications are Gold, 60% are Silver and 10% are Bronze. Also, a P1 incident for a Gold application has a standard resolve time of 4 hrs with an expected SLA of 99% and a minimum of 96.6%. This means that the supplier agrees to resolve P1 incidents for Gold applications in 4 hrs or less 99% of the time.

The data elements in Table 14-2 will vary for every support engagement due to the environment and nature of the applications involved. The SLA targets drive supplier costs. More labor is required to deliver SLA targets in which 70% of all incidents must be resolved in 4 hrs than it takes to resolve 30% of all incidents in 4hrs. I can't emphasize enough how important the SLA targets are. The SoW will also define default penalties for missing SLA targets. The supplier may face a financial penalty if the SLA results are expected and/or fall below the minimum. Getting this right is foundational to a successful relationship with the supplier and the ability to generate optimal value from the agreement.

Problem Management & Root Cause Analysis usually contains two measures. The time it takes to submit an initial analysis describing the problem with recommended corrective action. The second measure is the time it takes to submit the final analysis and implement the corrective action. Restoring service means the user can use the application to perform their work. This could involve a temporary "work-around" until a permanent "fix" (code change) can be installed. As noted in Chart 14-2, the standards for these metrics are also based on the tier level of the application (Gold, Silver, Bronze) and the severity of the incident.

➢ Data for the three preceding metrics can be found in the ticket database of the Call Management Center (CMC) and Help Desk (HD). Each ticket receives multiple time stamps as it moves through the system: time opened; time acknowledged; time/date resolved; and time it is fixed permanently. The database associated with the CMC and HD contains a wealth of data points that be interpreted to measure performance and improve the service delivery process.

Measuring performance is an important element of Supplier Governance and it will require resources to manage, maintain and report results. First, there is a tremendous amount of effort required to initially establish the tier level and incident criticality definitions for each application. Don't underestimate the time this will take to establish! Also, these definitions will need to be updated as new applications are added over the term of the contract. It takes time to gather the data and create consistent reports that easily paint the picture of supplier delivery. Ownership for managing the reporting process will be a part the Governance team responsibilities. The three key aspects of reporting include data collection; report formats; periodic review and follow-up on actions taken.

Data collection will encompass a number of sources. The most prominent being the database associated with the Call Management Center and Help Desk. The data should be structured to include time stamps and data values capture when the initial incident was logged; the ticket number associated with the incident; the specific application associated with the incident; time the ticket is acknowledged by the Support Team; the application tier level and criticality of the incident; time service is restored; a description of how the incident was resolved; the time that root-cause was determined; description of root-cause; the time the permanent fix was installed. The collection of this data relies on down loads from the CMC/HD database. The supplier can perform the activities involved with pulling the data, however, someone from the Governance team needs to review the data to assure that there are no obvious errors. Also, the supplier will identify "extenuating" circumstances that may be the cause of certain failed measures. For example, although the application was down longer than the standard SLA, it was the result of a damaged server and totally out of the hands of the application Support Team. Decisions will need to be made for each of these extenuating events to determine if the supplier's claim is valid and if the data for that incident will be taken out of the report.

➤ The data source to measure "User Satisfaction" will come from surveys that users are asked to complete.

➤ Data sources for "Uptime" come from multiple sources. Actual downtime due to incidents can be found in the CMC/HD database. However, downtime associated with routine maintenance may need to be posted to the report manually.

The supplier may track and record data for other metrics, for example Minor Enhancements (ME). In many support contracts, there is an agreed amount of time the supplier will dedicate to minor enhancements each month. The unpredictable nature of application failure means that some days, the Support Team spends extra hours addressing incidents. Other days they may have some "free time." To keep the team busy during their "free time" the supplier may agree to a predetermined number of hours (typically 10% of the total monthly work effort) each month to work on minor enhancements. Keeping track of the ME request list; priorities; ME's in progress, and ME's completed is usually managed on a manual spreadsheet.

Report formats should be easy to read. The first rule of reporting is the fact that management really doesn't concern themselves with past events. Of course they may never tell you that – management wants solutions. Design the report to show key results on ONE PAGE so it can be reviewed quickly. Use graphics to identify results that are trending above and below norm. Including a trend chart is also a good way to present a time-series of multiple reporting periods.

Table 14-3 is an example of a typical "top-line" SLA report. This report is created monthly for management review and shows the number of incidents for each Tier and Priority Levels within. Although this report is a roll-up of all applications, the underlying data exists at a level of detail that will support further analysis by the project team to identify root cause. Over time, as "fixes" are made to the environment and application code, the report should reflect reductions in total ticket count, percent resolved on time, and customer satisfaction.

Table 14-3 SLA Report

Service Level	Priority	Customer Tickets	Target Resolve Time	Baseline Resolved on Time	Actual Resolved On Time %	Cusomter Satisfaction Survey	Survey Count	Open Incidents Past SLA	Auto Generated Tickets	Supplier Resolved %
GOLD	1	200	4 hrs	99%	99%	88%	3	0	15	
	2	320	8 hrs	88%	89%	84%	10	40	615	
	3	345	16 hrs	65%	83%	77%	15	79	420	
	4	40	2 weeks	94%	93%			3	826	
Total		905					28	122	1876	96%
SILVER	1	10	8 hrs	97%	100%	75%	1	0	36	
	2	549	21 hrs	92%	96%	79%	11	41	340	
	3	3247	1 week	97%	96%	82%	70	210	481	
	4	222	3 weeks	99%	100%	80%	2	30	141	
Total		4,028					84	281	996	48%
BRONZE	1	1	12 hrs	93%	0%	0%	0	0	4	
	2	160	42 hrs	92%	96%	74%	3	42	135	
	3	841	2 weeks	98%	98%	83%	24	250	119	
	4	60	4 weeks	98%	100%	0%	0	214	29	
Total		1,062					27	506	287	50%

Data for SLA metrics are captured in the ticket management system database.

The pace at which reports are presented will vary depending on the audience and type of report.

On a weekly basis, the project team should meet to review results at the most basic level of detail. Applications don't get fixed from the top. Improvements can't be made without knowing what is happening at the atomic level.

On a monthly basis, the working team should meet with the project team to review results (Table 14-3) and discuss chronic and/or emerging issues that need to be resolved to continue improving the process. The data presentation in Table 14-3 is adequate for this discussion along with specific events raised by the project team.

Each quarter, the executive steering team should meet with the working team to understand major trends, be informed of potential issues, and review innovative ideas – many may have been suggested by the supplier.

Penalizing the supplier with default fees for missing SLA's is quite common in application support engagements. Although I must admit that in the past 15 years working with top tier suppliers, I don't recall ever collecting one. That may be a testament to the quality of the suppliers used, or the fact that the supplier was able to "earn back" the default. Default penalties can appear to be complicated so I will spend a few moments to explain them.

First, it is important to define the operating criteria that will trigger a default. There are no standard default criteria in use for all situations – the client and supplier need to define this in the SoW. The governance team is in a good position to monitor the default process and also assure the supplier is dealt with fairly and in accord with the terms of the contract.

Missing "expected" or "minimum" SLA targets will trigger a default. Table 14-2 shows the expected and minimum targets for each priority level in each tier. The expected value approximates the mean of actual results. This can also be a dictated measure in the event historical data is not available. However, it is best to use standards that reflect process capability. The minimum SLA is a value that should never be violated. Ideally, all values would revolve about the mean and never fall below the minimum. Default criteria might be stated as: Missing the expected value 3 months in a row will result in a supplier default. Falling below the minimum at any point will result in a default.

➢ The SoW should define how much the supplier might have at risk in the event a default occurs. This can be anywhere from 10%-20% of the total monthly invoice. Third party teams show a strong sense of pride in delivering the SLA. Failing to make SLA is a big deal to them.

➢ Earn-back opportunities should also be defined in the SoW. Earn-back, as it implies, is an opportunity for the supplier to eliminate (or earn back) the default by achieving results that are better than SLA for some period of time. Every supplier can have a bad day and run into a sequence of issues that exhaust their ability to deliver on all SLA's. Forgiving a default if the supplier delivers SLA's at or above expected for three months in a row recognizes good performance and gives the supplier an incentive to improve.

From time to time, SLA's may need to be modified and/or new ones added to the agreement. Language describing how this is to be done should be incorporated into the MSA and also reflected in the SoW. The client and supplier must jointly agree to modify existing the SLA metrics. As discussed earlier, the supplier's cost is based on the SLA targets. When they change, their cost structure may change as well. Keep in mind that this is a two way street and easing SLA targets may reduce supplier costs possibly resulting in a cost reduction!

Development projects use different metrics than those used for support SoW's. The data associated with support work lends itself better to SLA measures and TQ charting due to the continuous flow of data over multiple years. On the other hand, development projects are comprised of various specific stages with logical "gates" or checkpoints between each stage. The stage gates are good places to capture metrics about development work.

"Milestone Achievement" is a good metric to track. Essentially it shows if the supplier reached a defined milestone on time. Milestone dates for each stage will be found in the project plan. For example, the plan may call for the "Requirements Definition" stage to be completed and signed off by the client on or before July 1; the "Design" stage should be completed and signed off by Sept 1; the "Construction" stage should be complete by Dec 1. As each stage is completed the date is recorded and compared to the current plan.

"Deviation to Schedule" is an interesting metric to track as very few projects are actually completed as originally planned. The business partner may be guilty of making scope changes (last minute changes to functionality) that result in deviations to the schedule. The supplier may cause deviations due to quality issues, resource shortages and inadequate testing.

"Defects per Line of Code" reflects the number of coding defects created during the building an application. Most defects will be resolved prior to initial testing; however, some slip through and may not be detected until later testing phases. Capturing the number of defects caught downstream is a good measure of the supplier's ability to manage their code building and testing processes. The measure is based on the number of defects detected divided by the number of lines of code produced.

"User Acceptance Testing" (UAT) may uncover coding errors, design flaws or an inadequate definition of the desired outcome. It is usually the case that the supplier is blamed for any negative feedback however, it is important to understand the root cause of the issue in order to take corrective action.

Operating Level Agreements (OLA's) are used when multiple vendors are working along common boundaries. OLA's assure that vendors work together to deliver desired outcomes for the client. In many cases, the application support supplier is different than the infrastructure support supplier. When an application stops working, it may not be immediately known if the application or the infrastructure (i.e. a server) is causing the outage. Let us assume there is an active P1 incident occurring with a four-hour standard for resolve time. The SLA clock starts when the incident is routed from the Call Management Center to the support supplier. However, resolution requires input from the infrastructure supplier. If the infrastructure supplier doesn't view the incident as a P1 it may not get the attention it needs to resolve it within the terms of the SLA. The problem arises because the two suppliers are operating independently when it comes to incident management.

OLA's attempt to address this "involvement" void. One way to assure the right level of involvement is to penalize all suppliers for any missed SLA's. This will help, but the push back from the suppliers will be "why should I be penalized when the other guy messes up?" The obvious answer to that is "why should the user suffer because the suppliers CANNOT work in a cooperative manner?" Another way is address the issue is to create inter-supplier OLA's. Supplier X will need to respond to any incident participation requests from Supplier A within 20 minutes with validation of fault within one hour (i.e. determine whether it is a server problem or not). Establishing the OLA's is more difficult after supplier contracts have been signed. The complexity increases exponentially as more suppliers are involved. There may be a supplier managing applications, another workstations, still another servers, DBA's etc.

Key Performance Indicators (KPI's) are also used to help the client measure and monitor results. The difference between KPI's and SLA's is that a KPI is not described in the contract and doesn't involve default penalties. KPI's are simply measures that the client, and sometimes the supplier, deem important. KPI's are often used to better understand specific process or performance issues. Tracking downtime frequency and duration for important applications may help identify chronic issues that need improvement. Examples of KPI's include:

➢ Number of user requests for a specific application – an indication that improvements may be needed.

➢ Service ratio shows how many incidents are resolved vs numbers of resources working on them. This may indicate staffing levels in need of adjustment – up or down.

➢ Fix Quality – how many code fixes are done right the first time.

➢ Resource turnover rates and/or average term of service per supplier resource. Insist on reviewing resumes of the resources the supplier will use on your project. From these one can calculate the average years of experience. Tracking this average over the term of the contract this will indicate whether the supplier is maintaining, improving or deteriorating the experience level of the resources deployed to your project.

KPI's may become SLA's upon the mutual agreement of the client and supplier. The process for transformation from KPI to SLA is typically described in the MSA and/or the SoW.

We have developed the business case; identified the supplier; determined the metrics needed and reached an agreement on contract details. It is time for the final "due diligence" review before signing the contract. It is a chance for the client and supplier to review all aspects of the engagement and make sure going forward is still a viable alternative. The next chapter will focus on the due diligence process.

Chapter 15:

Due Diligence Prior to Signing the Contract

One thing is certain when preparing large outsourcing projects is this – they take time! During that time things change and need to verified before the contract is signed. The final step before signing the contract is what is called "due diligence" or a final review of all the factors impacting project viability. Table 15-1 will be the basis for our discussion of due diligence throughout this chapter.

Table 15 - 1 Final Due Diligence

Strategic Concerns	Tactical Focus
Objectives of Due Diligence	It is the last chance to review contract specifics before the client and supplier sign it
Client Concerns	- Validate the business case - Employee impact analysis - Tier and criticality levels - All SLA's need to be defined - Project logistics - Staff transition teams - Name leadership team and key SME's for transition
Supplier Concerns	- Validate SLA metrics are achievable - New work elements added since RFP - Prepare resources needed for transition
Supplier and Client	- Agree to SLA dynamics - Fine tune scope of project - Business-as-Usual language - Connectivity and protocols in place - Disaster Recovery and BCP defined - Sarbanes-Oxley (SOX) testing

➤ The expected costs and benefits are clearer than they were early in the project when the assumptions were initially developed.

➤ The landscape of the IT department and objectives of the

enterprise come into focus helping us decide if the outsourcing project still makes sense given other projects and activities.

➤ The organization needs to make a commitment to staff the project with the resources needed for success.

The objective of the due diligence process is to give all parties an opportunity to review and validate assumptions made while preparing the project. Many assumptions were made – some may no longer be valid. As we work through due diligence, we look at concerns specific to the client, the supplier and those that are of concern to both.

Client concerns:

They can be numerous especially if it is their first major outsourcing engagement. Is the business case valid? How will our employees be impacted? Will the supplier care about the user as much as we do? How will the supplier be monitored? How will we work with a supplier 8,000 miles away? How will we manage and maintain control of the project?

Validating the business case is an important step. After all, it is the core reason for considering outsourcing in the first place. The client needs to re-run the cash flow model using the most recent cost updates. By this point in time, the client should have done a thorough analysis to understand exactly how much is being spent on "current state" based on the fully loaded resource cost.

The supplier will have provided their cost and ongoing staffing model. Included in this will be annual rate increases as well as annual cost reduction targets. Make sure the rate increases and productivity savings are incorporated into the model. Don't forget to include the cost of the ongoing client governance team as well – the supplier won't make reference to this.

The supplier will be sure to provide one-time costs for their portion of Knowledge Transfer. Be cautious with KT costs – the supplier will show only the cost of their resources during the KT process plus travel for offshore resources to come to the U.S. but that's half the picture. The client will need to continue paying incumbent resources for some time during KT – who else will the new supplier learn from? Paying for incumbent resources also needs to be considered especially if there is an "up-charge" from the incumbent supplier for this service.

Severance costs could be very large depending on HR policy and the length of service for those incumbent resources being eliminated. Specific individuals should be identified by this point along with what their actual severance payments will be.

The cost of establishing and maintaining offshore connectivity should be included in the final analysis. However, the costs are generally less of an issue than figuring out <u>how</u> to connect with offshore resources in a reliable and highly secure environment. Keep in mind that a major portion of the resource base working on the project will likely be 8,000 miles away. The link between them and the client location must be up all the time. Clients need to make sure their internal IT infrastructure team and the supplier's have this figured out before signing on the bottom line.

The impact to employee resources needs to be reviewed. The names of those impacted as well as a plan for communicating to them should be drafted. Don't leak information about who is on the list...especially before the contract is signed. These people have dedicated a part of their lives to the organization and deserve to be treated with dignity and respect. The fact they are being let go is a business decision – it is not because they are bad people! The list of those impacted should have been reviewed and approved by HR and Legal by this point.

Before signing the contract, it is important to define the tier level (Gold, Silver, Bronze) for each application or application cluster. Equally important is to identify the criticality level for various types of incidents that may occur. This exercise determines the level of response the user can expect when incidents occur and needs to be defined before the supplier can provide an accurate bid for the work. The cost to meet SLA if every application was Gold and every incident were a Priority 1 is not a realistic model.

> Further complicating the process is that the user will want every incident rated as critical – after all, they need each one for them to do their job. Additionally, the users are probably accustomed to the IT team working in an "all hands on deck" mode when incidents occur. Prioritizing the work will help assure that the most important incidents are resolved first. In a world where everything is a top priority, nothing really is.

The client governance team will monitor supplier performance using the SLA's discussed in the prior chapter. As noted earlier, the SLA cannot be established until the tier level for each application and incident severity levels are defined. SLA's and their associated performance criteria must be defined and agreed to prior to the contract being signed. SLA's are a major cost driver for the supplier and the key to performance quality.

Logistics issues should be planned before the contract is signed. Someone needs to coordinate the matriculation of supplier resources into the client's office. Where will they sit? How many computer workstations are needed? Onshore resources typically use client assets especially when it is important to adhere to their workstation image. They will also need to be provided with secure access ability and, where needed, an ID card to enter the facility. This doesn't happen for free and the cost of it should be captured as part of the one-time charges.

Members of the transition team should be identified by now so they can hit the ground running once the contract is signed. Gaining agreement on who will be moved to the transition team and their availability may take a few weeks. It is also a good idea to start naming the subject matter experts (SME's) and the time commitment needed for them to train the new resources.

Supplier concerns:

The supplier will want to validate certain assumptions as well. The three main concerns are SLA's; gaps between the RFP and current discussions; and resource planning.

SLA's should be based on process capability. For example, if a specific application is only capable achieving an up-time metric of 90%, agreeing to a 99% up-time SLA is unrealistic. Regardless of what the supplier may tell you, they are not magicians. Agreeing to unrealistic SLA's might be a sign that they are desperate for the business and willing to say and do anything to get it.

➢ It is quite likely that there is not sufficient data available to provide an accurate statistical evaluation. It may take several months to collect enough data to accurately define the SLA. In these cases, the supplier and client may agree on everything about the SLA except for the target values and set the target after four to six months of data collection.

As noted earlier, it may take a few weeks to work through all of the details that arise from the point when the RFP is awarded to the final draft of the contract. During this time it is not unusual for new work elements to be identified and some to be removed from the proposal. The supplier needs to make sure that changes to the scope of the initial RFP are incorporated into their project plans and cost calculations. The project team needs to make sure these changes are reflected in the contract as well.

Once the contract is signed, the client will expect supplier resources to be engaged the following day. This is not realistic given the time required to create a transition plan; secure work VISA's; plan travel; find housing; etc. The supplier would like to start as soon as possible so they can start billing for services. Top tier suppliers will anticipate contract approval and start to address staffing activities in advance so they can begin engaging as soon as possible after signing. Also, large suppliers have the ability to utilizing resources working on other U.S. based projects to form a nucleus to build from. This is not only good for the client but also good for the career development.

Joint Client and Supplier concerns:

There are a number of topics the supplier and client need to address and arrive at mutually beneficial solutions. I have said it a lot, but can't say enough how important SLA's are to the success of the relationship. Agreement must be reached on the specific SLA's that will be used to monitor performance and the actions they will drive.

➢ Both parties must agree on SLA targets for "expected" and "minimum" acceptable results.

➢ Details about default penalties must be worked out and should include the maximum amount at risk (say 15% of the total monthly bill). The terms for the supplier to earn back the penalty by over-achieving results should also be defined.

➢ Reviewing historical support data will validate (or invalidate) SLA targets. A plan needs to be worked out when the data is nonexistent or doesn't provide clear statistical validation.

➢ A mutual understanding of the project scope and anticipated deliverables can be much more difficult than it seems. Take the time to list all "in-scope" applications, technology platforms, and outcomes expected.

Remember, even though English may be a common language, when working with foreign companies it is important to understand how "analogies" (especially sports analogies) and local vernacular may not translate well. My advice is to restate important points in a variety of ways to assure that they are properly understood. Testing for understanding is an important skill to develop when working with foreign cultures.

Business-as-Usual is an important concept common to application support projects. The intention is to make sure the new supplier performs all of the activities preformed by the incumbent resources without the need for an exhaustive detailed list. Major work elements are easily defined in the contract while Business as-Usual addresses all those little things that are literally too numerous to count but nonetheless essential to operations. By agreeing to follow the Business-as-Usual language, a new supplier is essentially saying they will do everything performed by the incumbent crew. The new supplier will need to invest some time during due diligence to thoroughly understand exactly what Business-as-Usual involves.

When outsourcing to an offshore supplier, a significant portion of the team may be half-a-world away. Making sure they are technically connected is important, especially for support work. There are a number of technologies available to connect the U.S. to the offshore team members. The final selection should be a team decision with input from the supplier; IT security; IT infrastructure and the IT project team. It is important to consider lead times required to order and install additional equipment and software in the project plan. Any of the one-time charges to purchase and setup the equipment and software should be reflected in the cash flow analysis along with ongoing costs for required maintenance and monthly fees to lease high-speed data transmission lines.

Disaster Recovery (DR) and Business Continuity Planning: Every organization that relies on IT capabilities in the day-to-day execution of it's business must be able to react swiftly when a disaster disrupts their data management and transaction processing abilities. When a major disaster strikes taking out entire data centers, the company can be severely crippled and possibly put out of business. In addition to the tragic loss of life, a number of firms impacted by the 9/11 terrorist attacks never recovered when their data centers were wiped out – they lacked a back-up plan or relied on a solution too close in proximity. For example, a data center in the Pentagon and the back-up site across the hall were both destroyed.

DR is a common term used to describe how the organization will reestablish its physical environment, transactions and data structures. The reality is that it takes time to recover these processes. The Business Continuity Plan describes how the business will operate until IT services are fully restored.

If we were able to take a cross sectional view of a data center, it would show servers, applications, transactions and historical data. Example: XYZ Stores is a retail business with 100 stores across the U.S. headquartered in Los Angeles, Ca. Assume a major earthquake wipes out their data center; the cross-sectional view shows:

➢ Cash registers in the each store will typically run off a local server but need to upload transaction events at the end of the day. With no data center, there are no uploads and soon the ability to run at store level stops as the local server storage space is consumed.

➢ Revenue in the form of cash should be fine as it is sitting physically in the cash registers. However, records of sales paid for with credit cards, debit cards, gift cards, etc. will be lost. Swift recovery is important as the banks may refuse paying on credit card transactions if remittance does not occur within few days of the customer purchase.

> Visibility to all inventory records is lost including on-hand counts in each store; shipments in transit to the stores; orders planned to ship from their DC's; and new inventory coming from the manufacturers. Without a view to inventory the planning process is compromised and the stores will soon run out of product.

> With no connection to the Sales Database, returns processing will not be possible. The Sales Database keeps all sales transaction on it so a customer can receive an accurate credit in the event of a return.

Once in DR mode, it is important to recover quickly in order to minimize financial exposure. Best-of-breed IT organizations can move operations to a remote data center and resume operations after a few hours. To effectively accomplish this, a complete replication of the physical hardware (servers, work stations, etc.), software applications, middleware, historical data and access to active transactions is required. Maintaining redundant environments is costly and requires extensive planning and testing.

For a rapid role swap to occur (moving to the back-up location) all hardware operating systems in both locations need to be at the same release level, the same is true for the applications. All transactions and data need to be available to the back-up site. Active transactions and historical data are often "mirrored" to maintain accurate replication. Practice makes perfect and the role swap to the remote site should be done frequently to make sure it goes smoothly if/when it is actually needed. The supplier will need to know the specific actions they will perform to assure recovery happens quickly and accurately.

> As important as DR is to the client's operations, don't overlook the fact that your supplier needs a solid DR plan for their side of the handshake as well. One of my clients was using satellite technology to track their fleet of over-the-road trucks. We visited the vendor to see their DR process. They operated a remote data center 300 miles from the primary site that was physically identical – right down to the server racks and desks. They even had redundant satellites in case one was damaged. In the event of travel disruption, they had a number of Humvees gassed up with off road directions to travel between the data centers. They practiced the role swap every month and were able to resume full operation in 20 minutes. The cost of losing the revenue as well as the disruption to their customers easily justified the cost of their Disaster Recovery solution.

Business Continuity Planning involves IT and the supplier but should be designed and managed by the business users. In our example of the XYZ Stores, Business Continuity Plans are basically a definition of how key processes would be executed manually in the event of a technical disaster. For example, inventory planning may require that someone physically count inventory, calculate requirements by hand and use the phone to communicate with the manufacturers. Credit Card sales may need to be suspended or use manual imprints (remember those?) to be remitted at a later date. The bottom line is that every key process, especially those in the revenue generation stream, will need a Business Continuity contingency plan. The business will push back and try to make Business Continuity Planning an IT responsibility – don't allow that. They need to design the process, train the users on what to do, determine how many people are required to execute and test it out to make sure it is as complete as possible.

The Sarbanes-Oxley Act of 2002 was designed to protect investors from the type of corporate accounting scandals that took down Enron, WorldCom and others leaving shareholders with nothing. There are a number of required "internal controls" defined by Sarbanes-Oxley that public companies need to validate in order to be compliant. Some accountants show concern when told their internal Sarbanes-Oxley controls will be managed and tested by a third party IT partner. However, in my experience, the top partner firms are more aware of the controls than are the typical IT employees. Also, they have solid processes in place to assure that compliance is achieved without issue. During due diligence it is important that the client and supplier fully understand each other's roles with respect to Sarbanes-Oxley testing.

The due diligence process can take a few weeks depending on the size to the project. It is critical that both parties agree to all roles and contract language. Equally important is to make sure the cash flow model remains attractive and management is committed to supplying resources. Once these are accomplished, it will be time to celebrate – you have earned it!

In the next section we will discuss implementing the project and spend some time understanding the cultural differences to be faced with outsourcing to an Indian partner.

Section 4:

Implementation

Section 4 will focus on implementation. After developing a successful business case, preparing the organization, and signing the contract with the supplier, it is time to implement the project. The working team has done most of the work to this point but needs additional resources now to manage the rigors of Knowledge Transfer and implementation. This phase will be visible to the entire organization and expectations will be very high. Having done a great job to this point counts for nothing if the project is not launched and executed to plan. All eyes will be watching!!

We will examine the roles and responsibilities of both the client and supplier relative to implementation. Knowledge Transfer (KT) is all about moving the body of knowledge from the incumbent to the new resource and must be done right. Since KT for application support differs from development work – we will examine each separately. Both require the project to be woven seamlessly into the organizational fabric, technical environment and infrastructure. Finally, we will address how to manage ongoing governance and discuss cultural differences and similarities one will experience when outsourcing to Indian firms.

Areas of Focus: *"The Inside of Outsourcing"*

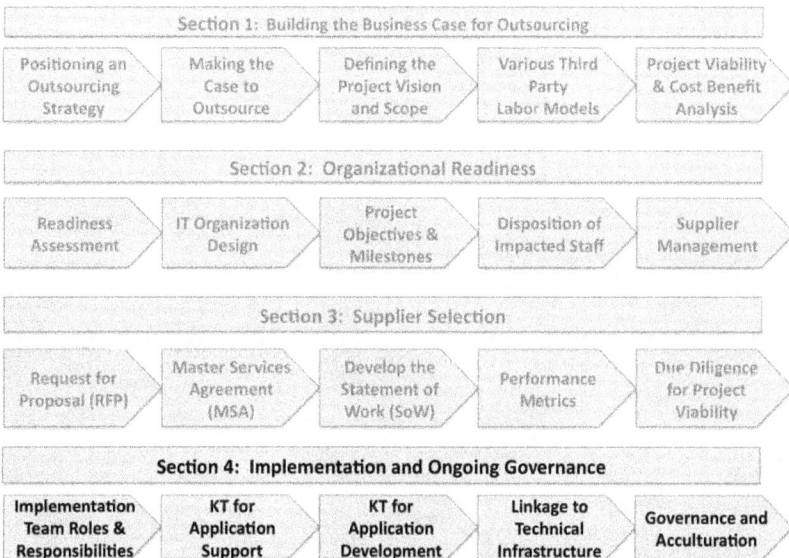

Section 1: Building the Business Case for Outsourcing				
Positioning an Outsourcing Strategy	Making the Case to Outsource	Defining the Project Vision and Scope	Various Third Party Labor Models	Project Viability & Cost Benefit Analysis

Section 2: Organizational Readiness				
Readiness Assessment	IT Organization Design	Project Objectives & Milestones	Disposition of Impacted Staff	Supplier Management

Section 3: Supplier Selection				
Request for Proposal (RFP)	Master Services Agreement (MSA)	Develop the Statement of Work (SoW)	Performance Metrics	Due Diligence for Project Viability

Section 4: Implementation and Ongoing Governance				
Implementation Team Roles & Responsibilities	KT for Application Support	KT for Application Development	Linkage to Technical Infrastructure	Governance and Acculturation

The *Inside* of *Outsourcing*

Chapter 16:

Implementation Team Roles & Responsibilities

Implementation is a team sport. To win, each member of the team needs to know the overall objective and what their specific individual roles are to achieve that objective. Following the outline in Table 16-1, we find there are four primary structures to understand: the client Working Team; the Process Team; the Supplier's on-shore team as well as their offshore team.

Table 16 - 1 Implementation Team Roles & Responsibilities

Strategic Concerns	Tactical Focus
Objectives	- Assure client and supplier resources know their role - Manage the organizational cultural changes associated with the transition to Managed Services
Team Structures	- Leadership, extended staff, supplier, business partners - Key characteristics and reporting structure - Identify direct vs adjunct resources for transition work - Onshore and offshore linkage and communication
Team Responsibilities	**Working Team:** - Understand project scope and contract details - Finalize objectives, timelines and detailed project plans - Monitor results and review milestone status - Manage impacted staff; communicate plan to organization **Process Team:** - Execute day-to-day knowledge transition process - Responsible for incident management during transition - Escalate issues with personnel, process, language, etc. **Supplier Team:** - Learn their new roles and take ownership - Provide application and/or technical expertise
Interactions Between Client and Supplier Teams	- Establish common goals and objectives - Define contact protocol between supplier, client and user - Determine required access levels and data security - Manage connectivity between onshore and offshore office

The overall objective of course is to plan the work, and work the plan in order to assure a successful go-live. Success can be measured in many ways but the three basic legs on the "success stool" include delivering the project on time, on budget and achieving the desired quality level. Conventional project management says that if any of the legs are changed, at least one of the remaining two will be compromised. For example, if the project finds itself behind schedule, it may take more money and/or a reduction in functionality to finish on time. If the project is costing more than it should, it may be necessary to reduce functionality and/or delay completion. If new functionality is added, it may take longer and/or cost more to complete the work.

But, contemporary wisdom is for losers! When the project goes off plan, it is time to refocus energy on where the breakdown occurred and creatively figure out how to remove the roadblock and restore the plan. Every project runs into snags...deal with them and stop whining about needing more money, more time, etc.

The Working Team (WT) earns its title during implementation, as there is an exhausting amount of work to be done. The leadership of the Working Team consists of the transition executive and three to four direct reports plus representatives from HR, Legal and Finance. Every aspect of implementation will default to the Working Team to resolve.

There are two primary objectives. The first is to assure that everyone involved understands his or her roles, responsibilities, time lines, etc. in order to deliver a successful "go live" for the project. This includes coordination with other parts of IT and business partners. The second objective is to make sure the organization adopts the Managed Services Model (MSM).

To be successful, a fundamental understanding of project deliverables, scope, and the contract are required. Early on I suggested the Working Team (including Legal, HR and Finance) be kept whole from project inception through implementation. Having been involved with projects in which one group creates the business plan and develops supplier contract only to toss it over the wall to a totally different group to implement. Failure rates are higher with this model along with a lot of finger pointing. I have found that when one group owns the project from soup to nuts, they pay a lot more attention to details when they know they need to actually deliver the goods! Its called "having skin in the game."

Equally important is a successful transition to the Managed Services Model (MSM) as it will be the foundation for future initiatives. At times, this can be more difficult than delivering the project itself. Changing the culture and behaviors needed to adopt Managed

Services is difficult. As with any culture change, the organizational "elasticity" will want to revert back to where it was. At risk is not realizing the full benefit of the project – I call this value leakage. The supplier must be held accountable for their contractual commitments. For example, in a support project the incident management processes must meet or beat SLA targets. For development work, it is all about meeting project milestones and code quality. Let the supplier resources resolve issues on their own.

It is common for retained client resources to help the supplier solve the problem quickly or worse yet, solve it for them. There are three key reasons for this. First, the retained staff is searching for a way to add value in this odd world of offshore outsourcing – it is new to them.

Second, a number of their co-workers were probably impacted. To avoid the same fate they work hard to show their value.

Third, they are afraid to suffer the wrath of their business partners and will do anything to please them. One way to do that is by jumping in to solve issues.

Although these sound innocent on the surface, they can lead to issues that inhibit the supplier from truly taking ownership. They may be tempted to deliberately undermine the new supplier with the business users as well.

The Working Team needs to focus on training the supplier, reviewing results and transforming themselves from "doers" to "reviewers." Working Team leadership must show their teams how their jobs have changed and what their new roles are. The extended team needs to be reassured that they still have a vital role and are needed for the long haul. Many suppliers offer workshops to help with the transition to the Managed Services Model. In some ways, the transition is similar to raising children. Parents want to be involved with every decision their kids make. However, their role should be to prepare them so that some day, they can make their own decisions. The same is true with MSM. Clients need to allow the supplier resources to make decisions on their own and evaluate the quality of those decisions with SLA's.

Table 16-2 is a sample matrix outlining responsibilities for various aspects of the transition. The Working Team co-creates this matrix with the supplier covering major areas of work from logistical needs such as workstations; establishing connectivity between onshore and offshore; security; and scheduling training sessions for Knowledge Transfer. This matrix forms the basis for the reporting structure of the implementation team. The supplier and client teams parallel each other. All individuals with two asterisks are Level 2 and report directly to the Transition Executive.

The first section of Table 16-2 addresses the broad foundational needs of the project and not related to specific applications. These roles are overarching in nature and independent from the project deliverables per se. Since they can support a broad range of future projects, I have them reporting to the transition executive. Many of these roles can be eliminated, or at the very least diminished, once the process matures and becomes part of the culture. Some roles will continue after the project reaches steady state with the resources moved into governance roles.

Most of the roles in Table 16-2 are self-explanatory however, "process integration" is important enough to expand upon. Process integration assures the client and supplier work practices mesh. The supplier needs to learn how to integrate themselves into various client functions such as change control, Sarbanes-Oxley testing, project management, security, etc. The client needs to learn how to mesh with supplier activities such as SLA reporting, offshore communications, code management, enhancements, etc.

The Level 3 individuals (no asterisk) in the bottom section of Table 16-2 are Process Team resources responsible for the transformation of each cluster. These resources report into the Level 2 leaders and many will remain after go-live for ongoing governance activities.

Table 16-2 Transition Team Roles

Transition Planning & Staffing	Supplier	Client
General Activities and Roles		
Transition Executive Leadership	Bill/Dennis*	John*
Contract Management	Nad	John/Tony
Engagement Staffing	Anil	NA
Process Integration	Sanjay**	Tim**
Logistics (workstations, id's, space planning, etc.)	Ravi	Ken
Infrastructure (Offshore connectivity, security, etc.)	Ravi	Jorge
Organization Transition/Communications	Gary	Rod
Work with incumbent to assure participation	NA	Jorge**
Application Knowledge Transition Cluster Leads		
Transition Program Leaders	Sri**	Paul**
Corporate Systems Group	Nisha**	Bruce**
Administrative Systems Cluster Leaders	Sanjay	Terry
Finance / Revenue Accounting Cluster Leaders	Nisha	Donna
HR and Employee Benefits Cluster Leaders	Balasu	Jake
Real Estate Systems Cluster Leaders	Balasu	Jake
Logistics & Inventory Systems Group	Srini**	Paul**
Transportation Cluster Leaders	Vijay	Kay
Warehouse Management Systems Cluster Leaders	Vijay	Kay
Inventory Management Cluster Leaders	Atul	Srina
Planning and Forecasting Cluster Leaders	Atul	Steve
Merchandising Systems Group	Karanta**	Barry**
Allocation and Replenishment Cluster Leaders	Suneel	Sally
Pricing Cluster Leaders	Suneel	Steve
Sourcing Cluster Leaders	Sudhir	Sally
Order management Cluster Leaders	Suneel	Sally

* = Level 1, ** = Level 2, no stars = level 3

Supplier resources mirror the client structure with key leads matched up between the two organizations. Team leadership is the single most important element behind successful transformations. The right staff will grow into strong champions of change and respected leaders in the organization.

You win and lose by the leaders you choose. Leadership characteristics to look for include:

➢ Visible ownership, understanding and energy needed to make the transformation successful. The staff has to be project champions and show their enthusiasm across the organization.

➢ Willingness to leave their comfort zones, trust the vision and demonstrate a tireless pursuit of excellence

➢ Ability to innovate, negotiate and motivate. Offshore outsourcing may be a new adventure for your organization, one without a roadmap to follow. Innovative individuals that can actually negotiate their needs with others in the organization are both vital and rare. They will become examples to follow and will motivate others.

Additional Subject Matter Experts (SME's) are needed to train and transition knowledge to the new supplier. The SME's will be loaned from other parts of the organization and many will only be needed for a few weeks going back to their primary roles. Do not assume the SME's are sitting around waiting for you to call. Work with their managers ahead of time to assure they can be made available as needed during the KT. To help their managers plan, try to quantify the number of hours per day and dates the SME's will be needed. Please note: If the vendor doesn't have access to the SME, timelines will slip, as he/she cannot learn the job – scheduling and availability is critical.

Table 16-2 also shows the suppliers' roles within the project and the structural pairing between supplier and client. These relationships are key to success and should be nurtured through cultural training, team building, and shared objectives. Relationship building is easy for those working in the same onshore office. But every effort must be taken to build a relationship with offshore team members too. I am a strong advocate of spending a few days each year visiting team at their offshore offices. This is the best way to strengthen the relationship and build loyalty with offshore staff. Be sure to budget the money and make these trips mandatory each year.

In order to successfully deliver the project, there are a number of specific roles that need to be filled and actions that need to be taken. The WT is responsible for delivering the project – period! To do that successfully they need to make sure all levels of the implementation team are functioning and working towards their objectives.

The WT needs to be the focal point for any questions that may arise regarding the project. To do this they need to:

➢ Understand the breadth and depth of the project.

➢ Articulate and communicate the vision for the project across their immediate teams as well as the broader organization.

➢ Assure the scope of the project stays constant by managing "scope creep" by not allowing the project to get bigger than agreed.

➢ Guard against value leakage by getting all that was promised out of the relationship.

In order to accomplish their goals the Working Team must have a solid understanding of the contract. Questions about pricing; staffing; which work activities belong to the supplier vs. the client; SLA's; change management and other aspects of the contract arise everyday during the implementation. Not understanding the contract will result in delays, under-delivered value, or worse yet, having to deal with the legal gang.

The WT defines the major planning components of the project and monitors progress to assure delivery stays on time and on budget. Objectives need to be clearly written and shared across all members of the team – client and partners. Working with the supplier, a detailed timeline of events and milestones is created and monitored.

A disciplined project manager needs to be named to track each element of the training schedule to assure that project timelines stay on track and don't slip. This person should report on progress each week and have authority to declare a "schedule breakdown" whenever the schedule begins to deviate from plan. Deviations develop whenever training sessions are missed; SME's are not available as planned; a gap in the transition is identified or an application was overlooked.

Detailed training plans are needed for each application and include: a named trainer (SME); the supplier resource to be trained; time of day and day of the week for training to be done. This level of detail takes time to assemble but is critical in order to keep the KT on plan. Most suppliers are the experts when it comes to creating the transition plan. Don't be afraid to use them.

One of the more unpleasant tasks is to notify the impacted staff that their jobs have been eliminated. I have always felt this message comes best from the immediate supervisor rather than pushing this task off to HR. However, HR should certainly participate in this sensitive topic but the manager needs to deliver the news.

The Working Team keeps the broader organization, including business partners, updated with general project plans, status and issues that arise. The Process Team (PT) does the "work of the work" and is very close to the actual knowledge exchange. Examples of Process Team roles include:

➢ Managing the day-to-day transition of knowledge from the incumbent resource to the supplier. This work is the backbone of a successful project and must be done well.

➢ Reverse Testing – after the initial training is completed, resources need to be tested for how well they have absorbed the knowledge needed to execute their new role.

➢ The length of transition for an application support project can easily span two to three months. During this time incidents will still occur and the Process Team will need to address them. They should use these incidents as live opportunities to train. It may be wise to suspend some of the non-essential activities during transition such as minor enhancements – there are only so many hours in a day.

➢ The Process Team is working directly with the new supplier's staff and will be able to spot any inconsistencies in skill level and/or language proficiency. They should raise any concerns they have to their leadership in order to assure the supplier is providing the agreed upon level of expertise.

The supplier will place resources onshore working along side the client Process Team staff with a much larger contingent of resources working offshore. Supplier roles include:

➢ Learning the job and taking ownership of assigned activities. Much will be learned through training, the rest through observation.

➢ Regarding the building of knowledge and the documentation captured in the Knowledge repository; make sure the contract stipulates that the client owns all captured knowledge and the repository itself.

➢ Provide application and/or technical expertise required to resolve incidents and generate improvements to the code.

➢ Deliver the inherent value of the contract. These include quality solutions that improve stability; minor enhancements that improve the functional experience of the user; and innovative ideas that lift the general status of the enterprise.

The interaction between client and supplier teams is the glue that binds the project together. Without a relationship built on trust and mutual respect for each other's capabilities, there can be no team. If there is no team, there can be no victory. Let's attempt to better understand the peer-to-peer relationships; interactions with users; access restrictions and technical connectivity.

Peer-to-peer relationships are shown in Table 16-2. For every activity or cluster there is a named resource from the client organization along with a counterpart from the supplier. These leaders should create and follow common goals and objectives for their respective levels. The WT leadership can provide a general vision, but tactics and goals to achieve that vision should come from the execution layer. It is a good idea to invest in team building activities that strengthen the bond and also bridge the culture gap.

Business Partner contact changes when support work is moved to a centralized Support Team. Earlier we discussed how the old IT organization was functionally organized with both support and development activities performed within the same team. Prior to centralizing, these groups created their own standards, work processes, supplier relationships and took the liberty to create their own IT strategy with the Business Partners – often independent from any overarching corporate strategy or IT vision. Centralizing support into an independent group provides a unique focus resulting in better continuity and efficiency across all support activities. However, as is the case with most organization changes, new dynamics require discussion on role alignment.

In the old world, the leader of each IT group managed Business Partner contact. In the new world of centralized support, there is often confusion about who owns that contact – the support lead or the development lead. Initially I could not understand why this was such a hot button but the plain and simple truth is this: Business Partner contact is power and people hate to lose power.

The change in organization structure results in the need for a new Business Partner contact model. The development leads believe they should own the contact role but running every incident through them is inefficient. Not only is there a time delay, those outside the Support Team simply don't have the information at their fingertips to effectively communicate the issue. On the other hand, Support Teams may not be in the best position to help the Business Partners create a long-term application strategy that may include replacement some day.

After managing a few transitions it dawned on me that "one size does not fit all" as each application is at different point in its lifecycle. In addition to work tasks, the "owner" of the Business Partner relationship should be based on where the application is in that lifecycle. We discussed the Application Lifecycle in Chapter 7 and I have replicated Table 7-2 below.

> As the application matures there are fewer and fewer dollars invested in it for new functionality. At some point the application is labeled "legacy" meaning it will be supported but no new investment will be made for additional functionality. As you can see in the lifecycle chart below, the work tasks assigned to the Application Support Team increase as the application reaches maturity. This is logical and allows the development resources to be repurposed to work on next generation application design and construction.

> Referring to Table 7-2 below, the Application Lifecycle concept we touched on in Chapter 7 can help create a new relationship model for each application rather than using a "one size fits all" approach. Since each application may be at a different place in its lifecycle, the model allows for a unique relationship for each major application in the suite.

Table 7-2 Application Lifecycle – Roles & Responsibilities

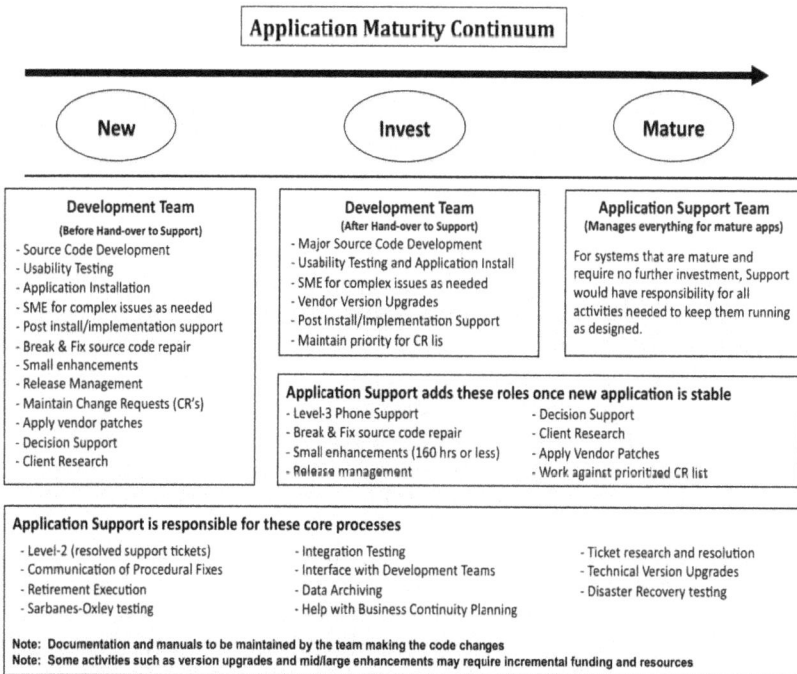

Application Maturity Continuum

New   Invest   Mature

Development Team
(Before Hand-over to Support)
- Source Code Development
- Usability Testing
- Application Installation
- SME for complex issues as needed
- Post install/implementation support
- Break & Fix source code repair
- Small enhancements
- Release Management
- Maintain Change Requests (CR's)
- Apply vendor patches
- Decision Support
- Client Research

Development Team
(After Hand-over to Support)
- Major Source Code Development
- Usability Testing and Application Install
- SME for complex issues as needed
- Vendor Version Upgrades
- Post Install/Implementation Support
- Maintain priority for CR lis

Application Support Team
(Manages everything for mature apps)
For systems that are mature and require no further investment, Support would have responsibility for all activities needed to keep them running as designed.

Application Support adds these roles once new application is stable
- Level-3 Phone Support
- Break & Fix source code repair
- Small enhancements (160 hrs or less)
- Release management
- Decision Support
- Client Research
- Apply Vendor Patches
- Work against prioritized CR list

Application Support is responsible for these core processes
- Level-2 (resolved support tickets)
- Communication of Procedural Fixes
- Retirement Execution
- Sarbanes-Oxley testing
- Integration Testing
- Interface with Development Teams
- Data Archiving
- Help with Business Continuity Planning
- Ticket research and resolution
- Technical Version Upgrades
- Disaster Recovery testing

Note: Documentation and manuals to be maintained by the team making the code changes
Note: Some activities such as version upgrades and mid/large enhancements may require incremental funding and resources

➢ The relationship model will change over time as the application moves through its lifecycle.

➢ When the application is new, even though there are support tasks, the primary "owner" of the relationship with the Business Partner is the Development Team.

➢ When the application is in the invest stage both teams must work closely together and share the relationship; both should be present at important meetings. The "leader" is the one who owns the topic being discussed. If it is a support issue, Application Support is the leader. If it is a development issue, the Development Team is the leader of the discussion.

➢ When the application is mature, the Support Team owns the relationship and performs all maintenance and non-discretionary development – such as updating tax codes, or employee benefits.

➢ It is important to realize the lifecycle status of an application continues to change over time.

Firms generally have good controls for security and data access. Theft of Intellectual Property is a risk that can be mitigated by background checks, access records, etc. One area that may present a risk is access to functional components of an application. It is easy to allow or deny access to an application suite. It is more difficult to control access to individual modules within the application.

Applications are often comprised of multiple modules of functionality. For example, the three-way match process in an accounts receivable application involves validating the invoice, verifying receipt of goods and authorizing payment. Separation of duties is critical and should be top of mind to make sure that access to various application levels is maintained for the supplier resources as well. Imagine the exposure a firm would face if the one individual had access to all three functions. That person could create an invoice from a phony company, verify receipt (of nothing) and authorize a payment to the phony company. Access restrictions and security protocols are intended to prevent fraudulent acts by controlling and recording access levels. Much of this is dictated by Sarbanes-Oxley legislation along with good procedural discipline.

➢ For a large corporation, small deviations are nearly impossible to detect. One of my clients had an employee that would create a $5,000 invoice payable to a fictitious firm that he had set up. He would then validate the invoice with a receipt of goods authorizing payment to his phony company. He had also set up a bank account to receive the funds. This guy was doing two transactions a day for 50 weeks and over the course of the year bilked the company out of $2,500,000. Since the average daily invoice processing of the firm was $10-$20 Million, a $5,000 deviation flew under the radar and went unnoticed. What is interesting is that an alert payroll clerk noticed a pattern of multiple $5,000 transactions in every week except two – the two weeks he was on vacation!! Further investigation revealed the scam. I guess the perp did not believe in paid vacations….

Connectivity is the lifeline between offshore resources and those working onshore. It is needed to keep onshore and offshore working together and when it breaks, approximately 70% of the team will be in the dark. Creating the technical linkages between onshore and offshore is part of the implementation process and can be managed by the supplier and the firm's IT infrastructure team. Plan for adequate bandwidth and make sure the link is tested before go-live.

In the next few chapters we will review the implementation process in detail for the two primary types of outsourcing projects: Application Support, and Application Development.

The *Inside* of *Outsourcing*

Chapter 17:

Knowledge Transfer for Support Projects

The objective of Chapter 17 is to define the elements of Knowledge Transfer for an application support project. Outsourcing support is different than development, so they will be reviewed separately.

Table 17-1 shows the five components of KT. It starts by creating a plan and ends with a successful go-live and approved Operations Manual that describes supplier responsibilities and processes. At the risk of being redundant, I am repeating some points in this chapter to create a consolidated planning guide for the practitioner.

Table 17 - 1 Knowledge Transfer Application Support Projects

Strategic Concerns	Tactical Focus
Objectives	- Understand and execute the various phases of Knowledge Transfer and make sure the new supplier is accountable
Project Planning	- Implementation approach - Identify resources and establish timetable & milestones - Gather transition materials - Communication to the enterprise
Knowledge Transition	- Transfer existing knowledge - Compliance to schedule - Assure successful transition
Secondary Support	- New supplier resource in back seat - Gain knowledge by doing - Co-locate trainer and trainee
Primary Support	- New resource is in the driver's seat - Incumbent is in the back seat - Co-locate for questions, suggestions, etc.
Steady State	- New supplier takes ownership - Incumbent resources are removed - Start using performance metrics - Monitor contract commitments

Project Planning:

As with most things in life, good planning is the key to success. This is especially so with outsourcing. Without a detailed plan to follow, the project will simply drift along and be subject to changes in organizational focus before it can be completed. Someone needs to be assigned the role of project "task master" with responsibility for tracking and reporting daily progress against the plan. This person needs the ear of the client and supplier project executives. A development project is a unique isolated event – if it fails it fails and another project and/or supplier takes over. The loss is limited to the money spent and time lost. Planning the implementation of an application support project requires a great deal more coordination than does an application development project. A support project generally spans multiple years and as a result of the outsourcing effort, a large number of knowledgeable employees will be exited. The loss of knowledge creates high risk that could cripple an organization if the project eventually fails.

The foundation for the successful implementation of a support project starts with the detailed work plan. Questions about sequencing, work clustering, timelines, etc. must be addressed in order to construct the plan. Plan on at least three to four weeks to create the detailed work plan. One point I would like to stress when creating the plan is to be cognizant of the seasonality of applications. For example, there is a certain time of year the employee benefits enrollment takes place. The KT for this cannot occur except during the enrollment period. If enrollment is in August, starting KT in October means the new resources will miss it completely. Year-end processing of Financial Applications is another example. One option is to bring the supplier in earlier than what the contract calls for to learn and shadow during these seasonal events. It will facilitate a faster KT with less risk.

There are a couple of ways to sequence the Knowledge Transition (KT) work. One is to arrange the applications into logical clusters as discussed earlier and rollout the clusters in time-sequenced waves. Another option for rollout is to use the aptly named "big bang" approach and implement them all with the same go-live date. Just be careful that the "big bang" isn't the sound of the door slamming behind you!

Typically, clusters are formed around common functional areas: Finance has its cluster; Inventory Planning has its cluster, etc. However, clustering can also be based on technical platforms: all mainframe applications form a cluster; midrange based applications form a cluster; etc. This is OK from a technologist's perspective as it creates a depth of technical knowledge. However, it doesn't mesh well with business users – they are usually arranged by function instead of technology. There is a hybrid approach that structures onshore resources along functional lines while organizing offshore teams by the type of technology the application runs on. This supports the business users and also creates a depth of technical knowledge at the same time.

Large initiatives should consider using waves, as it is the best way to manage and balance the workload of limited staff. Take the time to strategize the optimal way to sequence the waves. One approach is to create a wave that includes a partial set of applications from multiple groups i.e. an application from Inventory Planning, one from Finance, one from HR, etc. This spreads the workload across multiple teams while ignoring the linkages between the remaining applications in the cluster. I prefer creating waves arranged by functional clusters and staggering the start dates within the wave by a week or two. Even though Table 8-2 was covered in Chapter 8, it does a good job showing the waves and staggered start times for each cluster.

Table 8-2 Implementation Plan Overview

Wave	Sub Waves	Application Cluster	Week 1	Week 2	Week 3	Week 4	Week 5	Week 6	Week 7	Week 8	Week 9	Week 10	Week 11	Week 12	Week 13	Week 14	Week 15	Week 16	Week 17	Week 18
1	1A	Procurement	K	K	K	K	K	K	K	S	S	S	S	P	P	P	P	C	C	C
	1A	Transportation	K	K	K	K	K	K	K	S	S	S	S	P	P	P	P	C	C	C
	1A	Administration	K	K	K	K	K	S	S	S	S	S	S	P	P	P	P	C	C	C
	1C	Time Clocks		K	K	K	K	S	S	S	S	S	S	P	P	P	P	C	C	C
2	2A	Allocation & Replenishment	K	K	K	K	K	K	S	S	S	S	S	P	P	P	P	P	C	C
	2A	Inventory	K	K	K	K	K	K	K	S	S	S	S	S	P	P	P	P	P	C
	2B	Warehouse Management		K	K	K	K	K	S	S	S	S	S	S	P	P	P	P	P	C
3	3A	Product Design	K	K	K	K	K	K	S	S	S	S	S	P	P	P	P	P	P	C
	3A	Financial Planning & Analysis	K	K	K	K	K	K	K	S	S	S	S	S	P	P	P	P	P	C
	3B	Revenue Forecasting		K	K	K	K	K	K	S	S	S	S	S	P	P	P	P	P	C
	3C	Cost Accounting			K	K	K	K	K	K	S	S	S	S	S	P	P	P	P	C
4	4A	Field Operations	K	K	K	K	K	K	K	S	S	S	S	S	P	P	P	P	P	P
	4A	HR Benefits, Payroll, etc	K	K	K	K	K	K	K	K	S	S	S	S	S	S	P	P	P	P

Key	
K	Knowledge Transfer
S	Secondary Support
P	Primary Support
C	Complete - Steady State

Once the clusters, waves and sequences have been determined, a timetable can be created. Implementation has four distinct phases. The length of each is subject to the complexity of the application. The Knowledge Transfer phase can last anywhere from four to eight weeks; Secondary Support four to six weeks; and Primary Support four to six weeks.

The transfer of knowledge from the incumbent to the new supplier takes place during Knowledge Transfer. Individual training sessions pairing the SME and the supplier resource need to be scheduled and monitored for completion. The schedule identifies the SME, the supplier resource, the key elements of the application, the day of the week and time of day the training will take place. It is critical that access to application SME's is coordinated with the SME's manager. It is rarely the case that every key knowledge resource will be made available to work on the KT effort. There is other work going on that the SME's are involved in. Having a detailed plan helps to explain to their managers when and for how long they will be needed. Coordination with third party SME's present different challenges especially if the third party is being replaced. This is another reason to make sure that every vendor contract includes verbiage about the transfer out process.

My advice earlier was to is to spend time creating the language for an eventual exit. Hopefully your supplier partnership will last forever, but if it doesn't things will become very ugly when you try to transition in a new supplier. The ousted supplier holds all the cards and can make it very expensive for you. In the even that you didn't prepare properly, here are some suggestions:

➢ Agree to pay them for their participation in the Knowledge Transfer process. Pay a modest premium if need be.

➢ Offer to give them work on other projects.

➢ Work out some sort of bonus structure to give them an incentive to participate in earnest. Maybe some sort of financial kicker if KT reaches steady state on time or even early.

➢ Don't flex your muscles and threaten legal action unless you really have a strong case. Remember, they hold the cards!

Training tools such as application documentation and historical event data are used to supplement the face-to-face training. Most firms have spotty (if any) documentation. One of the benefits of working with Tier 1 suppliers is how they significantly update and improve the documentation artifacts as they learn the applications. They will also create a Knowledge Management Database in which all changes to the application along with all incidents and how they were resolved. This is also known as support documentation as the information is used to help resolve similar incidents as well as training future resources. Make sure its noted in the contract that the data, physical database and updated documentation remains the property of the client.

Once the project plans and milestones have been set, it is up to the Working Team to make sure all stakeholders across the enterprise are informed about the general plan and their specific role in it. For example: the IT Infrastructure team needs to participate in setting up connectivity with offshore; IT Security will need to set up access ID's for the supplier resources; HR will need to make sure new onshore resources have a place to sit, and are provided building access; Business Partners will be curious about how their world will change and who they will contact when incidents arise.

The internal incumbent resources should be notified that their jobs are ending. Do not start the KT until these people have been informed. Be sure to explain their future as best as it is known (layoff, re-assignment, SME for KT, etc.) and what their options are.

Knowledge Management (KM) is the process that captures, retains and recalls specific attributes, actions and changes to the application suite. The Change Management Database (CMDb) becomes the cornerstone for KM including all changes to the physical environment as well. It is a central database that logs every change to the hardware, operating systems, applications, etc. Although the CMDb concept is great, the effort to create and maintain one is huge. Until the formal CMDb is in place, the support supplier should build its own application specific database. In order for it to be effective, it must be kept up-to-date. It becomes the source of retained knowledge, and used to train future resources, which is especially important if there is a change in supplier. Without a strong KM effort, the client risks becoming a "knowledge hostage" of the supplier. Consider creating a new role on the Employee Governance team to manage the KM process and assure it remains vital and up to date on an ongoing basis.

The detailed work plan is created before implementation starts. Expect to spend four to six weeks creating the detailed work plan for a large project of 200-300 resources. Top suppliers have tools to help with the creation of the implementation plan.

Knowledge Transition:

The Knowledge Transfer (KT) phase does exactly as its name would imply – it transfers knowledge from the incumbent resource to the new resources of the new supplier.

In order to do this, the transition plan we just developed needs to be staffed with Subject Matter Experts (SME's) to teach the new resource everything they know about the application. As mentioned earlier, the SME may come from various parts of the organization and can even be a resource of the incumbent supplier – it doesn't matter. The important thing is that they understand the intricacies of the application and various associated support activities. Searching for the SME and working with their managers to assure they are available to participate in scheduled KT sessions can be a daunting task.

Compliance to schedule is managed by the project "whip." The whip is needed to keep the work on track making sure that every training session takes place and, if missed, is quickly rescheduled.

Attending training sessions and completing each session doesn't necessarily mean the new resource is capable of supporting the application. Therefore a concept called "reverse KT" is used to test the knowledge and capabilities of the trainee. The SME, perhaps with the help of the business user, creates a list of questions and presents situations to the trainees to test their knowledge. Results are scored and if the trainee misses more than expected, parts of the KT session is repeated until they get it right. Expect to spend six to eight weeks on the Knowledge Transfer phase depending on the complexity of the applications involved.

Secondary Support:

In this phase the incumbent is still managing the support process with the new resource in the back seat looking over their shoulder. The new resource starts by learning and watching but quickly gains experience and knowledge. In this phase trainees might learn by actually resolving incidents. They will also learn other support processes such as routine application updates; technical upgrades to a new release level; annual calendar changes; year end application modifications for Employee Benefits, new tax law changes, etc.

By the end of this phase, the trainee should be in a position to take over all support activities from the incumbent. One of the key types of learning during this phase is absorbing "tacit" knowledge. Tacit knowledge is purely experiential and not found in a manual or book. It is also called "tribal knowledge" and is usually passed down from person to person. To benefit from tacit knowledge, it is critical that the trainee and trainer are co-located at the same workstation and requires passing a reverse KT test. This is not something that can be done remotely using the phone or video conferencing. As noted earlier, a hostile incumbent may not allow trainees into their facility. Therefore, when you draft support contracts, make sure and create verbiage to describe what will happen if a change in supplier is required. Co-location of trainee and trainer is a key element in this description. The time planned for Secondary Support is four to eight weeks.

Primary Support:

Primary Support is when the new resource takes over the lead role with the incumbent is in the back seat – just in case. This is the final test of the KT process to assure that all training objectives are met.

Although the incumbent is still "technically" responsible for support, the new resource should be performing as many activities as possible required to support the application. We touched on Business-As-Usual earlier in the book. Business-As-Usual are all those little known activities the Support Team does which need to be taken over by the new supplier. The new supplier really needs to make sure they know what all of these are but it may be difficult to actually experience them depending on where things fall on the calendar. For example, if transition takes place in April, it may be difficult for the trainee to experience a calendar year-end close since it won't happen for eight more months. The time required for this phase is four to six weeks. Steady State:

Steady State is the official culmination of the project. I also like to call it "go live" for obvious reasons. At this point the incumbent resources should be moved to their new role and the new supplier has officially taken. The new supplier is now ready to have their performance measured with the agreed to SLA targets. All aspects of support are now the responsibility of the new supplier. The client role will change to one of monitoring SLA's to assure compliance along with the occasional coaching and/or training that may be needed.

There will be a lot of anxious people wanting to hang on to the incumbent staff long past go live "just in case" something goes wrong. Although they would never come out and say it, the new supplier would like to keep the incumbent resources around as long as they can too – it gives them a little insurance. Incumbent resources should be removed from the process as soon as possible. It is important to get them redeployed to a different cost center or exited in order to meet the financial goals of the project. By this point the supplier as well as any lingering incumbents will be charging time to the project. It is important to establish ahead of time (when the contract is written) how delays in the removal of incumbents will be managed.

➢ To be fair, incumbent resources should be notified in advance of their roles ending. The incumbent supplier will need time to redeploy them to other projects – just make sure the exit is planned to occur at the end of their role in the KT process. The exit process MUST be covered and agreed to in the original contract. Otherwise, when the time comes, you will have no control over the resources needed to transition knowledge to the new supplier.

Periodic performance reviews, or quality checkpoints, should be scheduled to review compliance of SLA's, identify chronic issues and plans to remediate, delivery of contracted improvements such as minor enhancements. Review sessions should take place weekly at the Process Team level and monthly at the Working Team executive level.

The *Inside* of *Outsourcing*

Chapter 18:

Knowledge Transfer for Development Projects

Time frames, objectives and desired outcomes for development projects are quite different from those associated with support work. This is why I have chosen to discuss them separately. Table 18-1 is focused on Knowledge Transfer for development projects.

Table 18 - 1 Knowledge Transfer for Development Projects

Strategic Concerns	Tactical Focus
Objectives	- Build supplier expertise so they can deliver the desired outcomes that build business value
Technical Environments	- Learn the environment and inter-relationships - Assess unique situations - Build the SoW for desired outcomes
Business Process Flow	- Know what the business does and how they use IT tools - Identify business leads - Learn how the data flows
Desired Outcomes	- Describe the expected outcome and how to achieve it - Develop high-level work plans
Project Planning	- Gather requirements - Design solutions and create the project timetable
Skills and Resources	- Determine required skills - Estimate resource hours and onshore/offshore mix ratio
Construction	- Build the application - Using Agile Development with offshore developer teams - Test the solution - Train users and implement
Evaluate Results	- Meeting milestone dates - Quality expectations met - How much did the project cost

It is difficult to discuss Knowledge Transfer for development and not get into the topic of project management. There are many books devoted to the topic – this is not one of them. With that in mind, this chapter is over-simplified relative to project management.

Technical Environment:

The technical environment is a series of unique relationships between the application and hardware it runs on. For example, many firms use Oracle and run it on HP UNIX. However, each of these will be slightly different from the next because of the configuration, database platform, release levels, etc. To understand Oracle, and/or UNIX, is not enough. The situation requires knowledge of the unique nuances in the relationship between them.

Assuming one has a good knowledge of the basic application and experience with the platform, there are some ways to build an understanding of the environment: interview SME's; read existing documentation; and support the application. Utilizing members of the Support Team on development projects can be invaluable since they have a solid understanding of the inner workings of the application from supporting it. The support supplier can make a strong case to be used for development projects too.

➢ Using support staff for development work is a good way to capitalize on the knowledge they have learned. However, one must be cautious about utilizing that knowledge and avoid the pitfalls of "too many eggs in one basket." There are some options to consider other than using the support supplier for development work.

  ○ There is no reason the development vendor can't partner with the support vendor to provide insight and information for Knowledge Transfer. One must be careful the support vendor is not put in a position where they miss SLA. The support vendor may need to increase staff to help our for a while and should be compensated for that.

  ○ Just to be safe, there should be verbiage in the contract requiring the support supplier to help provide domain knowledge when the client asks for it. Be sure to write the development contract to include funds to backfill support resources so they can participate in the KT for the development project.

  ○ Earlier we discussed the notion of assigning both development and support activities to the Support Team for those applications that have moved into the mature stage of their lifecycle. For these applications "development" is typically done for compliance and/or release upgrades and I would support using the support supplier for these.

Unique situations and radical dependencies should not be overlooked. By this I mean that some applications and their associated data structures are stretched to the max with very little room for additional transaction volume. Adding new functionality without considering network performance and thoroughly testing it might very well result in major processing delays and contention issues that could severely impact business viability.

The best way for the supplier to build their knowledge base is to start by making small changes to certain applications. This is a safe way to build and test the knowledge level of the new development supplier. Gradually increase the size and complexity of the project until they reach proficiency.

It takes time to build a solid knowledge base, and because of that, the use of multiple development suppliers in the same domain creates a difficult situation – a domain is a group of applications supporting a specific business function. Earlier in the book we discussed the concept of bidding development work on a domain-by-domain basis. This means that an RFP would be structured so that all future development work in that domain would be awarded to the winning supplier. This is a lot more efficient than conducting open bidding for each project. It also allows the supplier to recoup some of their initial investment in knowledge acquisition. Make a thorough evaluation before selecting the supplier for a domain.

Remember, even if a supplier is assigned all future work in a domain, it is not a *financial* commitment. How big is "future work?" Is it one project that takes 1000 work hours to accomplish or is it 6 projects each needing 500 hours of work? It is impossible to accurately predict future demand, however, one can make a promise to assign all of that future work to the supplier that wins the bid in exchange for favorable rates. Once funding is approved for a new project, a Statement of Work (SoW) should be drafted. The SoW defines the desired outcomes of that project, required skills, timing and financial commitments.

Business Process Flow:

Good code is not developed in a vacuum. The supplier needs to understand how the business processes are integrated to the technical systems. It is important to know how the business uses the various IT tools they have. One way to understand the application from the user's perspective is to sit in their chair and do the work for a couple of days – this is invaluable! What the supplier needs to know is this; how will the desired change to the application impact work performance – both upstream and downstream. The process flow diagram shown in Table 18-2 can be helpful in determining both upstream and downstream impacts that may result from a change to the business process.

Every business process does something – "doing" words always end in "ing." If your job doesn't end with an "ing", by definition, you are not *doing* anything. A business process converts inputs into outputs by some sort of action taken. Inputs come from upstream providers, and outputs flow to customers downstream. Whenever inputs change, the process may be affected. Changes to inputs may also affect the quality or cost of the outputs as well. Every output of a process becomes an input to the next process in the chain.

Table 18-2 Business Process Flow Diagram

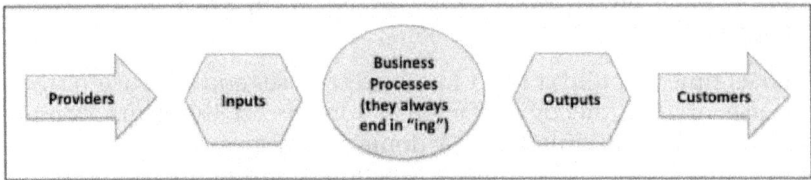

Providers	Inputs	Business Processes (they always end in "ing")	Outputs	Customers

The IT professional must understand the following to be effective:

➢ What the inputs are and where they come from.

➢ How the process converts inputs to outputs.

➢ The outputs and the customers that rely on them.

➢ How the process will be affected by changes to inputs.

➢ How to change the process to deliver different outputs.

In many ways, the business flow should mirror the data flow model needed to create successful solutions.

Desired Outcomes:

Application development resources must be an extension of the business team. A good working relationship with the business leaders and their staff is one of the fundamental requirements to building solid applications. It is critical for the supplier to understand the expected outcomes of the development effort. The SoW Managed Services projects should be focused on the outcome the project deliver to help the business user do their job better. This is different than the SoW for staff augmentation resources and requires some coaching so the rank and file understands it.

Knowing what the expected outcome is and how to achieve it

requires a great deal of input from the business partners. They know the work and how the current software operates as well as what needs to change. What they may not know is how to express these needs in "IT speak."

Table 18-2 is a simplistic example. In the real world, a business process is likely to have multiple inputs and multiple outputs resulting in a complex set of conditional combinations. It is difficult to test every conceivable set of conditions, which can lead to an unexpected error down the road. Testing on a complete replication of the production environment will help, but that, unfortunately, may be too costly to consider.

➤ The two questions on everyone's mind are "How long will it take?" and "What will it cost?" Experienced project managers can provide a reasonable ballpark estimate. However, without a detailed project design and a specific technical strategy, it is hard to be very accurate. I advise against providing premature estimates – these numbers are never forgotten and will haunt the project. It is preferred that a high level project plan be provided that outlines the major elements of work needed to finish the project. Once the funds are released a deep-dive can be performed to build a detailed work plan and cost estimate.

Project Plans:

There are six major phases common to just about every project. After each phase, approval should be granted before proceeding to the next phase. The first phase defines the concept and the desired outcomes associated with it. Clients will ask for the moon and if the supplier agrees to deliver it, the project runs the risk of becoming a black hole while expectations get out of whack. The supplier needs to help keep client requests reasonable.

The second phase is to determine whether or not the project is feasible and the outcomes can be delivered in an affordable manner. The third phase is project definition. This involves gathering requirements from the Business Partners. For IT development work resulting in a functional tool for the users, it makes logical sense to create something they can use to solve a business problem. Unfortunately, there are quite a few IT groups that "create software in search of a problem!"

Avoid designing the solution until all requirements are known. Keep the conversation focused on what is needed rather than what software to buy or build. Once all requirements are known, the IT project manager can present an array of potential technical options to the Business Partner lead. From there they can discuss and agree on the best solution to pursue.

Once the solution has been selected, its time to create the detailed design for the application. The level of detail should be sufficient to guide code creation and provide input for the project leader to develop a work plan.

Once the work plan has been established, the project timetable will be able to highlight major segments of the work associated with the development and delivery processes.

Skills and Resources:

Before proceeding with the final three phases (construction, testing and deployment) the project manager needs to make sure the skill sets and required resources are available. Many of the specialty skill sets, such as database analysts, will be in demand for other projects at the same time they are needed by your project. Don't make the mistake of assuming these folks will be available when you need them without scheduling them in advance with their managers. Other skill sets that need to be scheduled include: network architecture; security; spec writers; coders and testers. There may also be a need for resource expertise in those applications that the project will interface with.

Calculating the number of work hours required for each skill set is not only useful for scheduling, but will be important in developing a cost estimate for the project as well. This analysis should be done for each major phase of the project.

The onshore/offshore ratio of supplier resources will also determine the cost of the project. We spent considerable time on these in Chapter five and I will not go into that detail here. However, there are some instances such as "agile development" where a heavy use of offshore resources may not be realistic due to the nature of the beast.

Construction:

The final three phases of the project are construction, testing, and deploying the application across the user base. It is within these phases where the bulk of the project money is spent for both software and hardware. Construction involves developing the code that will become the application the Business Partners will use to solve a specific business problem and improve operating results. In most cases, the largest amounts of project funds are spent during this phase so proper planning and close monitoring will result in less waste and rework.

Agile Development is a group of methodologies based on using iterative incremental development. Requirements and solutions evolve through collaboration between cross-functional teams. Agile promotes adaptive planning and encourages rapid and flexible response to change. The Agile Development mindset is based on:

➢ Individuals and interactions over processes and tools.

➢ Working software over comprehensive documentation.

➢ Customer collaboration over contract negotiation.

➢ Responding to change over following a plan.

For those of you considering the Agile Development process, here are some observations that may impact project dynamics when using offshore resources.

Communication between onshore and offshore resource teams is vital to develop cohesiveness, provide a daily update, and address outstanding issues. A minimal shift overlap of a couple of hours should be considered even though it will be inconvenient for some or all of the team members. Remember, most of India is 11.5 hours ahead of Chicago. This means that when it is 8:30am in Chicago, the time in India is 8:00pm. If India starts their workday at 8:00 in the morning, the time in Chicago will be 8:30 in the evening. If the end of the traditional workday in India is 6:00pm, it will be 6:30 am in Chicago. Identifying a good time for a shift overlap of at least a couple of hours is difficult if not impossible without extending or creating off-hour shifts.

An online collaboration site is used to store specifications, test cases and active conversations. Avoid using email and enforce the need to keep all communications done via the collaboration site. This will become the permanent record and include all project artifacts including management tools for Agile Development.

Web conferencing should be used to create a sense of proximity for team members. They can be conducted on a daily basis to conduct reviews of specifications, and walk through requirements.

Ensure that the offshore team has its own development and test environments and code repository. Depending on where most of the development takes place will help decide which team (onshore or offshore) should have control of these environments.

Shorten the size of the development modules (sprints) in order to create more frequent inspections and provide better visibility to project progress for the client and the offshore partner. Shorter sprints will also make design changes easier and facilitate a more focused style of communication with the customer about project requirements resulting in quicker customer design sign-off.

With Agile Development, the main focal point for the project is the

Scrum Master. A scrum refers to the iterative process of moving the project forward – similar to how a rugby scrum moves the ball forward. Ideally the Scrum Master is fluent in the language of both the onshore and offshore resources creating better communication and understanding. Don't scrimp on the Scrum Master – assign this role to the best resource available!

The Business Partner (owner of the finished project) needs to define what "done" means for each stage of development. This is more difficult in an offshore model given the distance between locations. It is much easier to simply walk down the hall to refine and iterate. Along with well-defined requirements, acceptance criteria should also be created especially for the testing phase.

The offshore team must have all the requisite technical skills to be independent from the onshore team. A strong offshore team lead needs to manage the offshore staff and assure technical expertise is available as needed. The offshore team must be trained in the various aspects of Agile Development specifically in the implementation scrum. It is advisable to invest in training by bringing offshore to onshore (or vice versa) for a couple of sprints to ensure a common view of expectations and processes.

Quality Assurance (QA) testing should be done at the end of each major code module. There are several levels of testing and each should be successfully "passed" before moving on to the next module. Metrics for reporting the overall quality of the supplier's work are plentiful. The various levels of testing include:

➢ Unit testing – confirms the code segment actually works as designed. Any coding errors detected are corrected.

➢ Integration testing – confirms the code works when integrated into the applications or other major code segments it touches.

  o Note: Defects captured during Unit Testing and Integration testing can be reported in many ways but the most common is "defects per line of code." These metrics are straightforward and target values can be incorporated into the project SoW.

➢ Performance testing – confirms that the code delivers its designed output in an efficient manner. Failing to perform this test may result in major issues if the code, once installed, is unable to process the transaction volume. The root cause of defects captured during Performance Testing can be more difficult to trace with regard to cause. Sometimes the defect is caused by the code but it can also be caused by existing issues in the environment that come to light as a result of the project.

One of my clients built an application that tracked millions of inventory transactions. Performance testing uncovered an unacceptable response time for the user. Naturally the application was blamed. However, further examination discovered inefficient code within one of the databases accessed by the new application. The increased transaction volume created by the new application exposed the inefficient code in the database.

User Acceptance Testing (UAT) – as implied, the user actually confirms that the code performs in an acceptable manner and delivers the desired outcome required from the project. UAT is the final event in the QA testing process. If the user isn't happy, the project is a failure. It doesn't matter how efficient the code is or how pretty the screens are. The application must solve the user's initial problem, which is why UAT must be linked to the initial requirements gathered at the beginning of the project. Project artifacts must show the original requirements as well as any authorized changes to them. The QA team will develop a test plan to assure these requirements are met and tested. If UAT results show they were not delivered, the project gets a hit. If UAT shows that requirements are delivered, yet the user still isn't happy with it, perhaps the user takes the hit.

Evaluate Results:

Over the course of the project the client should monitor progress and evaluate results. When the project is delivered on time, with acceptable quality and cost parameters, the organization celebrates. Measuring whether or not milestones are achieved sounds simple enough, just measure actual completion dates vs planned. However, scope changes often occur and those coupled with other project dynamics can have an impact on the schedule – often these are beyond the control of the project manager. The clients may request additional functionality; management may decide to reduce budgets; emergencies may dictate that key project resources are pulled off to work on a higher priority issue. Unfortunately, the steel-trap minds of most corporate leaders will only remember the initial dates and want "heads to roll" when milestones are not achieved.

One method is to keep track of the initial milestone date and document any changes to it. The documentation should show the reason for the date change and who approved it. Tracking should show results vs the original milestone date and the revised date. The project team needs to manage changes to the schedule. The process should be spelled out well before the project starts so there is a clear understanding about how it will be managed and who is responsible for approving a change.

At the end of the day, it is important to know how much the project cost. Close to 90% of the cost of development projects is capitalized and generate depreciation expense. Most IT organizations use accounting tools to keep track of internal labor cost, third party costs and the cost of equipment. Resources from other IT cost centers (i.e. database analysts, infrastructure, security, etc) should also be captured.

All of these costs can be summarized and compared to the original estimated cost of the project. A monthly project budget report will show actual spending "to date" and compare it to what was planned. The report will also show the funding required to finish the project. Ideally the amount spent to date plus that required to finish does not exceed the original spending estimate.

The accuracy of reporting hinges on the requirement that IT project resources enter their time in labor tracking tools. Management reads these reports and takes action based on a perceived reality. Don't make the mistake of thinking labor reporting is done accurately – monitor it. Here are some typical responses from people when they were asked why their time reporting was not accurate: (hard to believe they are all college graduates!)

➤ "I didn't think I should report time until next year when the project will be done."

➤ "I didn't have any other work assigned so I charged my time to Project X so it didn't look like I was idle."

➤ "I was called into three or four other initiatives and just charged my entire time for the week to project X."

➤ And the classic: "My manager told me to charge time to Project X because his budget for Project Y (which I am working on) was overspent for the year."

The next chapter will focus on the linkage required between the project teams (both support and development) and other IT functions such as infrastructure, and security.

Chapter 19:

Linkages to the IT Environment

The technical environment of any major company is a conglomeration of servers, applications, telephony, vendors, users and processes that we often find connected in a manner resembling a plate of spaghetti – a mish-mash of technological touch-points. The success of any project hinges on its ability to recognize and effectively navigate within and between these elements or nodes in the environment. It is especially important to understand "who is responsible for what" (often called "roles and responsibilities") at the boundary where two or more nodes touch. Table 19-1 is a potpourri of the many touch-points to be examined in this chapter.

Table 19 - 1 Linkage to the Environment

Strategic Concerns	Tactical Focus
Objectives	- Evaluate the project's impact on the broader organization and understand how they weave together
Technical Linkages of Concern	- Hardware requirements - OLA with infrastructure supplier - Resolve timing conflicts - Source Code Management (SCM) - Coordination of major work efforts in progress
Roles and Responsibilities	- Incident management - Managing minor enhancements - Strategic plans for application suite - Who leads? Who follows? - Infrastructure upgrades
Working With Multiple Third Party Partners	- Replacing the incumbent supplier - Supplier gaps and overlaps - Technology vendors and IT consultants

The main point of this chapter is to create awareness and understand how a specific project will impact the broader organization – think of it as an "environmental impact study." Although not unique to third party relationships, the complexity of working with multiple stakeholders and vendors adds to the difficulty and importance of doing a thorough analysis. The project leader needs to identify all of the touch-points associated with the project – both physical and procedural. Physical touch-points represent connectivity to servers, other applications, databases, etc. Procedural touch-points are where the application and the user come together. New applications or major upgrades often require modifications to the way users do their job.

I have segmented our discussion into three major categories: technical linkages; linkages between application teams and business partners; and linkage to multiple third party partners. We will explore the physical and procedural linkages within these three areas.

Technical Linkages:

Understanding the technical environment and how a project relates to it is a critical exercise. I cannot identify any significant application or business process that is purely stand-alone. To the contrary, the environment in which most applications exist is a complex web of interrelated technologies and business processes.

The first question to answer when installing a new application is "What server will it run on?" Every application needs to run on something. The project team, working with the infrastructure group, must answer the following hardware related questions.

➤ Is there adequate capacity on an existing server or does the application require an increase in processing capacity?

➤ Where will the database reside? What is the anticipated transaction volume, number of records and frequency of puts and calls?

➤ If incremental servers are required, what is their cost and lead-time to deliver and set up?

➤ Will the project drive incremental labor costs in the data center?

➤ Is the bandwidth of existing telecomm technology adequate to manage the increased flow of data transactions and/or voice communications.

   o One of my clients decided to allow employees to access and utilize social media in the office. Response time slowed to a crawl because there was no investment in bandwidth to handle the data volume from streaming video, music, etc.

Operating Level Agreements (OLA's) are used when multiple suppliers overlap. The OLA spells out how they will work together to support the enterprise. Without an OLA, there is a risk that the supplier will only take action when the results benefit that supplier. This can be devastating to the organization when complex incidents occur. A properly written OLA will assure that suppliers work together to resolve incidents in which both are implicated. The ability to create effective OLA's starts when the supplier contracts are written. One thing I like to do is hold all suppliers accountable to common SLA's. Suppliers need to commit to working and acting like a team focused on the benefit of the user. When applications break, the user doesn't really care why they broke. They look to IT as the sole entity responsible for keeping their applications running. There is no value to the business that comes from finger pointing or assigning blame. The users look at IT as one big box of mixed nuts (literally and figuratively).

A broad view of the landscape across the enterprise will identify conflicting projects. The CIO's staff should have a good handle on all major IT initiatives. They need to manage the IT project plan to keep work flowing as smoothly as possible and within the bounds of available resources and approved spending limits. There may also be synergies between projects.

> One of my clients was very interested in outsourcing their application support work. At the same time, they were also engaged in a major Enterprise Resource Planning (ERP) implementation. Synergy came from repurposing outsourced resources from the Support Team to work on the ERP implementation. The results were quite impressive:

  o The cost of severance was minimized. The work on the ERP projected lasted two to three years. Normal attrition meant the affected resources did not need to be let go.

  o The cost of resources for the ERP project was reduced as well. Employees from the Support Team were able to use their unique knowledge of the applications being replaced to add immediate value to the ERP project. The internal employees cost about one fifth as much as the external ERP consultants.

"Source code" is the current version of an application running on a production server to process and/or generate transactions. Changes to existing source code and the creation of new functionality should be done in an isolated development environment. Once the new code has been completed it needs to be tested. If testing is successful, the changes can be moved to production. Documentation of the code change should be made and stored in the application archives. Authorization to make changes must be secured when the code is checked-out and again when the code is ready to "move to production." Even with thorough testing, there are instances where the new code simply doesn't work and needs to either be fixed on the fly or backed out of production. Knowing who made the change and reviewing associated documentation makes this unpleasant task easier.

Source Code Management (SCM) ensures authorized personnel are the only ones able to gain access and make changes to the source code. Good Source Code Management also prevents one developer from overwriting the code changes made by another. Interestingly, third party suppliers are often the ones to point out holes in the client's security protocols around Source Code Management. Imagine the problems that can occur when the code isn't locked down. Changes can be made to the source code that can cripple the functionality of the code. Not knowing who checked out the code or the nature of the change leaves the IT Support Team in the dark about what happened and how to fix it. Keeping a handle on the many environments in which code resides is a full time job that should be assigned to someone – usually someone in the infrastructure group.

➤ So what happens at 3:00am when there is a support issue that requires a minor code change? The pre-work and authorizations discussed above may not be possible during the off-shift hours. Yet, there may be users sitting idle waiting for the code to be fixed. The "Hot ID" concept is used for emergency code repairs. Those responsible for SCM create an authorized list of Support Team members. Code checkout is granted and the support staff makes the minor source code changes. Once completed it is moved back into production so the application can be reinitiated. Since this is done on an emergency basis, the documentation is completed after the code is back in production – usually next day.

Managing cross-functional projects is difficult. Projects generally require the use of specialized resources for specific periods of time. Specialists include architects, database analysts (DBA's), QA testers, etc. To optimize efficiency, most IT organizations have a "central pool" in which these experts reside. For example, the DBA's report into one group and deployed as needed to work on various projects across the enterprise. This allows the DBA's to grow their core technical expertise and knowledge of databases. Growing the technical expertise is more difficult and more costly if these resources were part of the staff of various IT functions. Without comprehensive resource planning, the availability of that specialty resource when needed is purely based on chance.

➢ Here is the problem – let's say that annually, there are about 20,000 hours of project work performed by our DBA team on multiple projects. In a perfect world, this amounts to ten individual resources. If the work is spread evenly across the year, resource availability is manageable. However, when demand for the DBA's is not smooth, problems can occur. Here is what can happen if 30% of the project demand for DBA's occur between May and June.

   ○ Each DBA produces 320 hours of work over two months. (40hrs per wk X 8 wks). Our ten DBA's can produce 3200 hours of work over two months. However, if 30% of the total annual demand for DBA's falls in May & June, those ten resources need to produce 6,000 hours of work. There are three options available: work each DBA for 15 hours a day (not likely for two months); hire more DBA's (not easy on short notice) or allocate the DBA's to the higher priority projects which will result in delays for other projects.

Effective cross-functional planning can alleviate the issue. However, the concept, although a great one, is still more of a concept and less of a reality. There are many vendors in the marketplace that will be happy to sell you their planning tools. Unfortunately, most of these tools require a level of input and commitment that is rarely available in any IT organization. Besides, even with a solid plan, there will always be that unexpected project that comes along and upsets the balance of nature. Flexibility is the answer to being able to quickly ramp up and ramp down the number of specialty resources in play. Large suppliers can help with this by utilizing the strength of their numbers to bring in DBA's, architects, etc. on short notice to help out. I am not advocating outsourcing the entire DBA group (or any other specialty for that matter) but rather using the supplier of IT services as a flexible resource pool that can quickly respond to changes in demand.

Roles and Responsibilities:

Cooperation between IT teams is important and here are a few examples where further work is needed to establish organizational protocol and a clear understanding of roles and responsibilities.

Incident Management is clearly the responsibility of the Support Team. However, questions arise about who should communicate and discuss support issues with the Business Partners. The Support Team is best suited for this since they are most familiar with the issues. Nothing is gained by asking the support staff to update the development staff so they can update the Business Partner – this is an example of organizational waste and redundancy.

Minor Enhancements are typically managed by the support team as well. Make sure the contract includes provisions for the supplier to provide a specific number of hours each month for Minor Enhancement. The number of hours is about 10% of the total monthly work effort. Minor Enhancements are small (less that 150 hours) and are not critical to how the application functions. They are modifications that will help make the user's job easier. The list of requests for Minor Enhancements will grow rapidly. Someone needs to be in charge of collecting and cataloging them. Next, the list needs to be prioritized so these enhancements can be built, tested and installed in order to optimize the user's capabilities.

➢ Don't make the mistake of trying to reduce the support budget by eliminating Minor Enhancements. Incident management, by its very nature, is not predictable and the work is something that can't easily be put off to tomorrow. Some days people are working 10-12 hour straight to resolve application incidents; and on other days they may have a little slack time. Minor Enhancements fill in the slack time. The only way to reduce the budget is to eliminate a percentage of the staff. With fewer people working on support, those 10-12 hour days become the norm resulting in burn out and eventually a degradation of service levels.

The IT organization should maintain the strategic plan for all applications. Investing in modifications and upgrades to current tools cannot go on forever and eventually new tools will replace the old ones. The Development Team should be responsible for working with the Business Partners to strategize next generation tools. Once an application is designated as legacy, it will no longer receive funding for major functional enhancements. However, because some legacy applications might be used for years before they are eventually eliminated, they will require upgrades for compliance. These upgrades keep the application current with tax codes, local laws, accounting standards, etc. The Support Team should perform all maintenance, and compliance upgrades, to keep legacy applications functioning.

Changes to responsibilities often follow an organization redesign. We discussed the benefits of centralizing support earlier in the book. Centralization provides opportunities for consistency and efficiency across all applications. To capture the benefit, it is important to make sure the primary teams (support, development and infrastructure) know their responsibilities in the new structure.

Table 19-2 is an example of a "Roles and Responsibility Matrix" that captures the responsibility for various roles. In reality the document may grow into many pages reflecting all touch points.

Team leaders must agree who "leads," who "participates" and who is "informed." The matrix becomes a valuable tool to help cement organizational and cultural changes associated with outsourcing.

Table 19-2 Roles & Responsibility Matrix

Topic	Development	Support	Infrastructure
Incident Management:			
• Manage the support desk	• Informed	• Lead	• Participate
• Notify Business Partners regarding major support issues	• Informed	• Lead	• Participate
• Modify code to resolve incidents	• Informed	• Lead	• Participate
Managing Minor Enhancements:			
• Collect requests from users	• Informed	• Lead	• Participate
• Prioritize list of enhancements	• Informed	• Lead	• Participate
• Work with users to build, test and install	• Informed	• Lead	• Participate
Application Strategy:			
• Work with Business Partners to design next generation	• Lead	• Informed	• Participate
• Build major new functionality for existing applications	• Lead	• Informed	• Participate
• Build, buy, install new applications	• Lead	• Informed	• Participate
• Maintain basic functionality of Legacy Applications	• Informed	• Lead	• Participate

Lead: Responsible for managing the action and making decisions
Participate: Provides input to decision makers
Informed: Follows up with execution of the action

Working With Multiple Suppliers:

As the use of third party partners expands, the complexity of the IT organization grows. Outsourcing IT services not only changes the organization structure, it requires that suppliers work together for the good of the client – and often this means some bending on their part. We will discuss the following four points in more detail: replacing an incumbent supplier with a new supplier, issues with supplier overlap, managing technology vendors, and dealing with IT consultants.

At some point there may be a need to replace an incumbent supplier. The reasons for this can be many and varied. Perhaps the incumbent is not able to perform as promised, or they are not capable of providing the skills and/or functional expertise the client requires. Whatever the reason may be, replacing an incumbent supplier is very costly especially if the incumbent has been around a long time and "owns" much of the subject matter expertise needed for knowledge transfer. In addition to the cost, the replacement of an incumbent supplier requires a great deal of effort on the part of the client. It is important to include some verbiage in the initial agreement rather than trying to retrofit the contract during the divorce. Every supplier contract should address a process for transitioning out. Most clients overlook this and end up paying dearly when the supplier needs to be replaced. The following examples of verbiage may help determine what to include in the contract:

➢ Note: All knowledge acquired by the incumbent supplier remains the property of the client IT organization. This includes all physical and electronic archives, intellectual property, software, databases and general documentation about the client's applications and associated environment.

The incumbent supplier should continue providing services they are being paid to perform until the client decides the new supplier is capable of taking over. When writing the contract, be sure the language requires the supplier to make key resources, named by the client, available for KT to a replacement supplier. Otherwise, these SME's can (and likely will) be redeployed to other customers. With that in mind, some sort of financial arrangement may be required to offset the suppliers lost opportunity and to assure KT is successful.

The incumbent supplier may have resources offshore. Be sure to include language that allows the new supplier to perform KT activities, as required, in the incumbent's offshore location. The new supplier must have the ability to meet with the offshore SME's.

Contract language will also be needed to assure the incumbent supplier works in a cooperative manner to help "transition in" the new supplier. The cost for this should be defined as well.

With multiple suppliers playing in the sandbox, there is a strong likelihood that there will be service gaps. Gaps would be specific aspects of the service package that have not been assigned. These must be addressed quickly once identified. At the same time, there may be overlap in other areas. Overlap is defined as two or more suppliers being assigned the same tasks. Overlap not only causes confusion it is costly – why pay two suppliers to do the same thing? As the client and suppliers fine-tune their working relationships, the overlaps should be addressed. First address the gaps to assure consistent service, then identify and eliminate overlaps.

The world is full of people trying to sell their software, hardware, technology and services. These vendors will go to great lengths to get their foot in the door. Some will drop names – "I was talking with your CIO last week and he suggested contacting you directly to take a look at our new product." Some will go "backdoor" and meet directly with the CEO to pitch their goods. It all sounds great to the CEO and they will pass the vendor along to the CIO who in turn will pas them along to you. The time required to deal with all of these vendors can overwhelm the organization. The best way to manage them is to establish strict policies for all to follow.

➢ Rule #1: Vendors are NOT allowed to pitch their product directly to the CEO or CIO. This of course requires that the CIO and CEO buy into this idea and support it. One way to sell upper management on it is to point out the need for the organization to avoid legal issues by treating all vendors in the same way with no special privileges.

➢ Rule #2: All vendors should be routed through the Supplier Relationship Management (SRM) team. We discussed the SRM concept at length in Chapter 10. SRM provides a vital filter to the IT organization that limits vendor access to respond to an expressed need for the type of product they are selling. SRM is an internal broker that matches resources in the marketplace to needs within IT. For SRM to work well, all contact with external entities must pass through them.

IT consultants should follow the same path described above for suppliers and vendors. As we discussed in Chapter 3, consultants can play a vital role in the IT environment. However, it is a good idea to establish ground rules for when to use consultants, as they can be quite pricy.

➢ Make sure the consultant has the expertise and experience needed for the work they will be hired to perform.

➢ Define the expected outcome from the engagement as clearly as possible – state what it is we want the consultant to provide.

➢ Establish a reasonable length of time the consultant will be needed. Open-ended agreements can be problematic. Generally, long-term engagements should use the Managed Services Model rather than a consulting arrangement.

➢ Consultants should be assigned to a client sponsor to assure they get the needed information to accomplish their task quickly and effectively.

The objective is to create a process to keep all suppliers of IT services working in harmony, similar to conducting an orchestra and NOT like herding cats. Each supplier, just like each musician (or cat), may be top notch in what they do. However, if they are ineffective at working together they will not perform well as a group. The role of the governance team is to monitor and manage the behavior and performance of the supplier long after the implementation is complete. We will examine governance in the next chapter.

The *Inside* of *Outsourcing*

Chapter 20:

Governance and Acculturation

Once the implementation is complete it is time to shift focus to managing relationship quality with the supplier on an ongoing basis. You will recall that we included governance resources in the cash flow model developed in Chapter 5. Without governance, your project will eventually fail because the supplier relationship will deteriorate. Table 20-1 is an outline for discussion to cover Governance Team objectives and specific roles; cultural assimilation and relationship management. The discussion is focused on application support since support contracts are multi-year in nature and require ongoing management. The outsourcing relationship for a development projects may end when the project is completed.

Table 20 - 1 Governance and Acculturation

Strategic Concerns	Tactical Focus
Objectives	- Create a structured approach to deliver ongoing oversight over the life of the contract
Governance Roles	- Compliance to contract - Validate SLA performance - Approve code changes - Liaison with Business Partners - Monitor/test resource capability - Assure processes are followed - Guide the change management process Move the relationship forward
Manage the Partnership	- Display partnership ideals - Don't use the contract as a club to beat the supplier - Resolve issues with supplier in an impartial manner - Recognize accomplishments - Address the critics
Acculturation	- Understand cultural differences; recognize similarities - Retain cultural identity while fitting into the U.S. culture - Body language and communication peculiarities

Governance Roles:

The objective of the Governance Team is to create a structured approach for the oversight of outsourcing initiatives. In the beginning of our journey, we created a Working Team and later added a Process Team to help facilitate the transition. The IT staff chosen for these teams was selected because of their expertise and knowledge of the applications and business practices. These same individuals should now be used to staff the Governance Team. They understand the project from its very inception and have worked hand-in-hand with the supplier. They want the efforts of the team to be seen as a success in the eyes of the IT organization.

The first question most executives will ask is "How many resources are needed to perform governance activities for application support?" A typical range is 10-20% of the total resource base associated with the outsourcing initiative. However, this number will vary by firm, complexity and also by the stage of maturity the project is in. This means that if we outsource support activities to 100 supplier resources a reasonable size for the Governance Team would be ten to twenty staff. Once resources mature and the project is stable, the number may be reduced – but never fully eliminated.

> ➢ Don't make the mistake of starting with fewer resources thinking you will be able to add more later if needed. Take advantage of the structural change that outsourcing provides and insist on getting the right number of resources you need.

Prepare to educate upper management on exactly what governance is; the roles the team will perform; and the number of resources needed to accomplish their mission – especially if this is the first major outsourcing initiative for your firm. The information contained in this chapter will help build your case. Management's perception of outsourcing is that a third party supplier is being hired to do all the work and the additional Governance Team resources are not needed and are nothing more than incremental cost. Change that perception!

The Governance Team will benefit by utilizing the very individuals that led the implementation. Organizing them by functional area, similar to how they were organized for implementation, optimizes effectiveness by capitalizing on their experience. The specific alignment of Governance Team resources may vary from firm to firm but should be structured to deal with all governance activities. Client employees, *and only client employees*, should be assigned to positions on the Governance Team. They will interact heavily with the supplier's resources to gather data, understand process dynamics, and create strategic plans.

The Governance Team monitors and maintains the health of the relationship between the client and supplier. They also need to validate that contract inclusions are delivered assuring that "value leakage" is minimized. Typical inclusions that need to be validated:

> Minor Enhancements – these are small changes to the code (usually about 10% of the total work effort per month) the supplier delivers during idle time when they are not working on active incidents. The "change request" list will grow quickly with priorities based on business need. The Governance Team must work with the Business Partners to establish and record the correct priority.

> Validate Sarbanes-Oxley compliance by monitoring procedures the supplier is to perform such as sign-offs on code changes.

> Code Quality is a metric showing the number of defects the supplier produces when code changes are required to resolve an application incident.

> Disaster Recovery testing can be a complex process. Typically a DR test is executed every couple of years to assure that in the event of a major data center calamity the firm can continue to function. Supplier resources are needed to participate in the test and validate that applications are functioning properly after a role swap has been completed.

> Validating SLA's will assure they are accurate. The Governance Team does this by sampling data to verify the supplier is achieving agreed upon service levels in the execution of their work. Comprehensive procedures are required to collect, monitor, store and retrieve the supporting data upon which the SLA's are calculated. This data can also be used to generate improvement ideas for the broad service delivery environment. The data will point out chronic issues that need attention as well as areas where positive changes have been implemented.

During the incident management process, minor changes to the code may be required. These changes are often discovered, coded and installed by the supplier after the client has reviewed and approved the change. The Governance Team must have the expertise to determine how proposed changes to the code will affect both the application as well as the business process supported.

> The supplier will identify technical improvements to the application and establish reviews with business owners. When possible, technical and functional changes should be bundled together thereby packaging up one code change instead of two. This improves efficiency but requires coordination across

multiple groups, which can best be performed by the Governance Team.

➢ The break-fix process may require an immediate code change in order to bring the affected application back to desired service levels. These changes need to be checked for completeness and require a Governance Team sign-off to assure they don't interfere with other parts of the code.

New development and/or changes to existing applications are a reality. It is the Governance Team who manages touch-points between support and development assuring both are constantly exchanging required information.

➢ Release plans should be provided by developers and include the date for the release; contact information for the installation team; the environments impacted; changes to batch schedules; how functionality will change; back-out plans and much more.

➢ The Support Team resources should be given documentation that includes Test plans and results; development documents; change logs; and rollback plans so they can be as prepared as possible to support the new code.

Liaison with Business Partners is a critical role for the Governance Team because the Business Partners are an integral part of the service delivery system. They need to be informed of incidents that disrupt the business – especially those not resolved within SLA. They should be updated about the root cause and the nature of any workaround solution that may require user training. The Business Partners should also be an integral part of designing the permanent fix. Users work with the applications daily and understand the functionality provided by the code. When a code change is considered, the Business Partners can quickly determine the impact it may have. This involvement not only helps assure the fix will work but also provides the Business Partners with the knowledge needed to train the users on new application functionality.

As more and more processing is added to the existing infrastructure, issues can occur during certain periods of the year when transaction volumes are at their peak. This requires that more attention be paid to data processing capabilities. The Governance Team will work with the Business Partners, Infrastructure and the application teams to plan for these events. The holiday season is when many retailers "make" their year. Black Friday (historically the day after Thanksgiving in the U.S.) is a huge day for retail in the U.S. as it signals the start of the peak holiday sales season. The last thing you want to do is to tell your CEO that sales were lost due to technical failures within IT...

Resource Capability determines the effectiveness of the outsourcing relationship. The Governance Team is in the best position to monitor and determine if there are any issues that need to be addressed. Supplier resources turn over from time to time and the supplier has responsibility to train replacements. For the most part they do a pretty good job with this but occasionally a bad resource slips through the filter. Perhaps their technical capability is lacking or their language skills are not sufficient. The Governance Team is in the best position to escalate issues with under-performers. It is important to bring any performance issues to the supplier quickly so they can take action. Top firms will work to correct the issue quickly and not allow it to persist.

Assuring that processes are followed is the key to successful management. Too often, management looks for "people failure" when things go wrong. I don't believe people intentionally want to do bad work. Bad results usually come from bad processes. The top outsourcing firms have well-defined processes. They will help the client acclimate to these processes in order to be successful. An example would be to make sure the user contacts the Call Management Center (CMC) to report an incident rather than calling the Support Team directly. When the CMC is bypassed, critical information is lost and the proper resource to address the incident may not be contacted when that person is 8,000 miles away in India!

There are many other critical processes that need to be followed and the Governance Team is in the best position to monitor and assure process compliance. Their over-arching vision of the environment includes many segments along the service delivery chain. Without holding firm, processes will eventually erode leading to confusion and critical value leakage.

Change Management captures all changes to the service delivery system. Server upgrades, new hardware, changes to peripheral devices, code change, etc. all need to be captured to maintain a current picture of the service environment. Occasionally a processor on a server will fry or some other physical component will fail – but that is rare. The overwhelming cause of application incidents is some change to the code or the technical environment. To quickly resolve the incident it is important for the support desk know what changed. The Knowledge Management Database should be the source for what has changed. It is a valuable tool that the supplier must keep current. The Governance Team should monitor it for completeness assuring that changes are posted to it in a timely manner.

Moving the relationship forward is done by working with the supplier to find innovative solutions to current problems as well as ways to preempt the occurrence of future problems. Top suppliers have a great wealth of experience and will utilize it to offer generic suggestions that fit the bounds of your specific business model. Sometimes the innovation is a new way to architect an existing application to make it more efficient. Sometimes it is an entirely new way of doing something that hasn't been done before such as pushing store specific events to the individual SMART phones of potential customers as they pass by. Not every idea is right for your business. The Governance Team needs to work with the supplier and Business Partners in order to determine what will add value to your business.

Manage the Partnership:

Building and maintaining a productive working relationship with an outsourcing partner is the key to long-term success. Remember, once the client decides to travel the path of outsourcing, it is difficult to undo that decision. Going back to "the good old days" is time consuming, costly and most likely will not be tolerated by upper management, as it requires hiring new staff along with a hefty price tag to train them. Maintaining partnership ideals when interacting with the supplier is a must to building the relationship in a positive direction. What follows are some tips and suggestions to help build and strengthen the relationship.

➤ Build trust by treating the supplier firmly yet fairly and be sure to establish the proper levels of linkage. Make sure that the right levels (client and supplier) meet to resolve issues. Nothing builds a relationship faster than solving a common problem together. Nothing destroys trust faster than one side escalating to a level higher than what is required or appropriate.

➤ Suppliers will naturally want to expand their business and in some cases that can be beneficial to both parties. However, one must evaluate their proposals with a critical eye. Don't be afraid to say "NO" and don't be afraid to say "YES" either.

➤ Keep things professional and work to avoid becoming so friendly that objectivity is lost.

One of the key roles of the Governance Team is to assure that business values within the contract are achieved. A solid understanding of what is in the contract and how the supplier will deliver it is needed.

➤ Caution: Be careful not to use the contract as a club to beat the supplier; it is really just a guideline for the working partnership. Unfortunately, I have seen occasions where one side or the other has difficulty expressing their concerns to the other and gives up. This breakdown in communication can lead to using the contract as a weapon losing the intent for why the two parties decided to work together in the first place. There is no way to draft a contract that will eliminate the need to interpret the intent of the parties. When the contract becomes a weapon, it is usually a two-edged sword and both sides end up bleeding.

Issues will arise. Get used to it. More important than the issue itself is how it is resolved. Issue resolution is more about understanding where the process broke down than it is pointing fingers at an individual. Focusing on the process instead of the people keeps emotion out of the exercise and leads to a more rapid solution.

➢ I have long been a fan of using data to understand and describe issues. Statistical Process Control techniques such as run charts, scatter diagrams, Pareto diagrams, etc. are tools that allow the data to describe the essence of the issue. People can react in various ways to what I may think or say, but they either have to accept or reject the data. The only way to logically reject the data is to provide alternative data.

  o As I have often told my students, the best way to lose an argument is to start by saying "I think..." Nobody really cares what you *think*. One is much better off by starting the argument by saying "The data shows..." This immediately shifts energy and focus to understanding the data and why it is the way it is. We must first understand what is happening before we can take actions to correct it.

The supplier is a valuable source of ideas that will improve the environment they are supporting. Engage them on a regular basis. Encourage "blue sky" or "imagineering" sessions to kick around improvement ideas that may lead to code changes or improved business processes. I have seen suggestions from these sessions lead to improvements that didn't previously exist in the realm of possibilities.

Be sure to recognize hard work and good results. A simple "thank you" made in a public setting has amazing power. A note of thanks to the individual and a cc to their manager is also very powerful. I would hold quarterly meetings with the entire staff (client and supplier) and key business partners to review and applaud accomplishments. We even arranged for remote offices and offshore teams join the meeting via conference call. Typical items to share with the group include:

➢ SLA metrics. Most of the time, the results were very good and fun to share. Sometimes the presentation included results that weren't so good – but they existed and had to be dealt with. I felt it was more important to present all the facts so people knew where to focus rather than paint a rosy picture.

➢ The challenges a specific team will face may be different from what other teams face. When recognizing team performance I was careful not to compare one team to another. I prefer to compare a team's results to what they accomplished in the past.

➢ Minor enhancements are a hot topic with the users. They were quite impressed with the number of enhancements and the speed of delivery.

➢ We also recognized personal milestones such as birthdays, service anniversaries and other items safe to share in public.

➢ Some of the business achievements included updates on transitions, development projects, general business results and other items of interest to the team.

➢ Team-building events can be fun and as the name indicates, can go a long way towards building the team. They don't always have to cost a lot of money either. If designed properly, they can tap the creative juices of the team and leave lasting memories.

  o In one of our meetings we divided the attendees into random teams and gave them used newspapers. Their object was to create the most elaborate hats.

  o Another event was to build the tallest tower out of Oreo cookies. The winning team got to eat the loser's cookies!

  o My favorite was when each team was assigned a movie title and had to perform a small skit demonstrating the theme. The group howled when Bruce came in riding on Rod's shoulders – both covered by a cape made from a tablecloth.

➢ It is important to reward achievements. Even in tough economic times, there are things that can be done to tell the team they are special. Potluck lunches cost little and create an interesting opportunity to sample foods from different cultures as well. Do something…anything!

Many people try to look tall by making others look small. The successes my teams have achieved often gained the praise of upper management. At the same time, there are forces at hand that will try to take any opportunity to discredit what was done. As noted earlier, the best way to win an argument is with data. Here are some tips to deal with your critics:

➢ Always be sure to invite them to open meetings and take the extra time to meet with them to get their input on future changes. Ask them to be your partner whenever possible.

> Keep emotions on the sideline using data to express your point of view. The critic will be forced to accept your data or provide their own. If they show their own, you can accept or dispute it.

> Flush them out. Often they will sit behind closed doors and with quiet voices, sow their seeds of discontent with anyone who will listen. Try to force the discussion out into an open forum. If they have a valid concern (and sometimes they do) ask them to voice it openly so corrective action can be taken. If their concerns are not founded in fact, they know they will look silly to the broader audience. It may cause them to shrivel up and crawl back under their rock.

Understanding the Cultural Divide:

Cultural Assimilation is formally defined as the adaptation of a minority culture to the ways and customs of the prevailing majority culture. Full assimilation results in an individual blending entirely into the new society losing most or all aspects of his or her previous cultural identity. Although this is the case for some Indians working in this country, most undergo "acculturation" or adapting to some aspects of the new culture while retaining much of their own native identity. While both assimilation and acculturation share a common process of adaptation, assimilation is much more extreme when compared to acculturation. Indian firms spend a great deal of time and money teaching their resources about how to adapt to the western world so they blend in more effectively.

It is important for U.S. firms engaged in outsourcing to learn and understand cultural differences in order to build a stronger relationship. It is also a great opportunity to learn about the Indian culture and customs. The richness that comes from learning about other cultures is not only good for business – it is good for the soul. Here are some observations that you may want to keep in mind as you move forward with your outsourcing initiative.

> During the time India was under British rule (from the 1700's to 1947) they had a great influence on Indian culture including drinking tea and driving on the wrong side of the road! Driving in India is done from the right side of the vehicle using the left side of the road – same as in the U.K. The following factors often lead to tragic consequences.

> In the U.S. driving is done from the left side of the vehicle and uses the right-hand side of the road. This is completely opposite of what Indians are used to. Having driven in the U.K., I can attest to the fact that it takes a bit of time to adjust to this change. For example, in the U.S., whether driving or

crossing the street, we naturally look for traffic coming from the left first, and then from the right. In India this is totally reversed for them and creates a few "near misses" (and sadly, some direct hits) while interacting with traffic in America.

➢ When driving in traffic, to avoid a head-on collision, we naturally swerve to the right in the U.S. The Indian driver would naturally swerve to the left – right into oncoming traffic!

➢ Driving at higher speeds than they are accustomed to can be a problem as well. Those of you who have traveled to India have witnessed the traffic congestion which, when combined with the absence of high-speed roadways, keep vehicle speeds relatively low. Not so in the U.S.

➢ An additional remnant of British rule is that Field Hockey is the national sport of India. In fact, India won Olympic gold in field hockey six times from 1928 through 1980. Cricket is a popular game in India too, along with most countries that, at one time or another, were part of the British Commonwealth. Indians are quite competitive when it comes to sports and games.

  o I had an Indian intern working for me that was determined to play on the office softball team. Sri spent a lot of time at the local batting cage to sharpen his batting skills. Having played a lot of cricket as a youth, hitting the ball came easy to him. Catching a fly ball? Not so much. He was in the outfield and ran up to catch a well-hit ball. Unfortunately for him, he misjudged the speed and trajectory of the ball and it struck him on the bridge of his nose breaking his sunglasses and causing severe physical (and emotional) pain. He was literally a bloody mess.

One of the more interesting aspects of Indian culture is the age-old custom of parents arranging marriages for their children. The parents find acceptable spouses for their children and introduce them to each other. At an appropriate age, and usually after some dating, the couple decides if they are ready to commit to marriage. Arranged marriages are still common but as social mores slowly change, the occurrence of "love marriages" is increasing. Love marriages are the result of two people meeting on their own and falling in love instead of allowing the relationship to be arranged by their parents. Although divorce happens, the rate is much lower in arranged marriages than in love marriages.

➢ One of the more interesting Indian customs associated with weddings is the *duration* of the celebration. It is not unusual for the wedding event to last many days running from early morning to late in the evening. It is an amazing event to attend if you ever get the chance. Be sure to bring a change of clothing!

In India, there are cultural mores that strictly prohibit public displays of affection between men and women. However, displays of affection between two males are a different story. One of the cultural norms is that men are comfortable holding each other's hand in public. I see this a lot in India. The first time I saw it my host reassured me they were not gay (not that there's anything wrong with that) but just good friends. One of the things Indian men are taught in their cultural assimilation class is that in the U.S. hand holding between men is not generally accepted and is reserved for gay couples.

Traditional Indian clothing is quite different from what we wear in the western world. Most Indian males working in the U.S. wear what their American counterparts wear. However, women are a little more uninhibited and are equally comfortable in their traditional saris as they are in the latest fashion from The Gap. The younger women tend to favor western fashion whereas the mature ladies are more likely to wear traditional garb.

Do not be offended if you take your Indian partners to your favorite steakhouse and they order pasta or some other vegetarian option. First of all, as you know, cows are sacred animals in India freely roaming the streets and neighborhoods. Therefore, the number of Indians that eat beef is very low. Many eat fish and chicken and some will eat lamb. About 30% to 40% of those I have worked with are strict vegetarians. A little creative forethought regarding where to hold business dinners will help minimize angst. The best cuisines to consider are Indian (duh), Chinese and Italian. All of these have an ample selection of vegetarian dishes on the menu.

Another interesting distinction is most Indians prefer their beverages without ice. It makes sense since ice and refrigeration is recently becoming more available across India. I have learned that it helps with digestion too!

I have found the education level of supplier staff working in the U.S. to be quite high. Most have college degrees and many have a Masters Degree. The third party resources I have worked with are very good at explaining what is going on deep within the coding layers of an application.

Many Indian nationals working in the U.S. want to move back to India to raise their families. They feel the primary education system in India is preferable to that found in the U.S. It is not a slam on school quality in the U.S. but rather a desire to have their kids grow up in the Indian culture and identify with their past. They also want their kids to grow up in the presence of extended families as well.

The constitution of India declares the right to freedom of religion similar to the U.S. Any specific religion can be practiced. The primary religions of India are Hinduism (80%), Islam (13%), Christianity (3%) and Sikhism (2%). Indian's celebrate a vast number of holidays during the year. Hinduism alone has 13 major holidays celebrated in India and across the globe as well. Two of my favorites are Holi and Diwali.

➢ Holi is a spring festival also called the Festival of Colors. During Holi, brightly colored paint and powder is thrown at everyone, everywhere. I saw the custom first-hand a few years ago and thought there was some sort of political demonstration going on but it was just people having fun. One of the interesting characteristics of Holi is that distinctions of caste, class, age, and gender are suspended during the celebration.

➢ Diwali is the Festival of Lights and celebrated in late fall. It is a New Year festival in the Vikrama calendar and falls on the night of the new moon in the month of Kartika. It lasts 5 days and consists of lights (similar to Christmas lights), fireworks, and lots of sweets and treats. Needless to say, it is a favorite with children.

Although there are many interesting differences in the Indian culture vs our own, I am more amazed by the similarities. Perhaps the similarities result from the fact that human evolution isn't very old by Earth standards resulting in genetics that are amazingly similar across the globe. This gives some basis to the theory that civilizations did not arise independently but rather from the cross fertilizations of nomadic tribes over the centuries.

➢ The importance of family is quite evident along with a great amount of respect shown to elder generations. Children are taught to be accepting and respectful from a very early age.

➢ Over the years I have befriended many Indian nationals and I must say the loyalty and openness they show to friendship is refreshing.

➢ The Indian resources I have worked with are aggressively dedicated to the job they are doing. They work long hours to assure the work is completed successfully. The resources I have had the pleasure of working with are top notch and dedicated to common business goals and truly interested in building a solid partnership.

➢ Another similarity is the common interest in finance and investing. There have been many interesting conversations with my Indian friends on broad economic issues as well as personal investment planning.

➢ The large Indian outsourcing firms tend to employ a high percentage of women – many at high levels. This not the result of political correctness or some government edict but rather, it reflects the high regard for the quality of work they do.

All things considered, the Indian resources working in this country are doing a good job fitting into *our* culture – much better than I would do if I had to work in India for an extended period of time. The geographic area they are working in can make a difference too as some are more easy to adapt to than others.

The climate in India is similar to that in Florida or Texas – hot and humid. Those assigned to work in colder climates have additional issues to face such as the fact that most of them have never owned cold weather clothing (jackets, scarves, gloves, etc) and many have never been exposed to snow and ice – especially driving on it.

Vegetarian diets are much easier to accommodate in large metropolitan areas. U.S. cities such as New York, Chicago and San Francisco have a number of vegetarian (as well as Indian) restaurants. These cities are also more likely to have an available supply of the spices used in Indian cooking. However, certain spice blends, such as Garam Masala and Curry are unique "family " blends handed down from parent to child over the generations. My friend Venky arranged for his mother to send spices to him on a regular basis.

Indians are also respectful of U.S. traditions and participate in many of our holidays. This is especially true for families with children in elementary school. It is hard for kids to go to school and abstain from participating in the annual Christmas Pageant, or exchanging cards on Valentine's Day or getting dressed up for Halloween.

I am impressed with the quality of their written word. The ability to clearly and concisely express their ideas on paper using impeccable English is top notch. However, with all cultures, communication peculiarities exist in both the spoken word and in body language.

➢ Body language is important to understand. For example, unlike western cultures, a head nod doesn't mean they agree with what you are saying. It simply means they "heard" what you were saying. This can lead to many arguments down the road. Be sure to ask, "Do you agree with me?" rather than assume.

➢ Similar to the above, "yes" doesn't always mean they are in agreement with you either. It may only mean they heard you. Again, be sure to ask rather than assume.

> ➢ Indians generally tend to be soft spoken and rarely raise their voices. I personally like people that are passionate about their work and confident in expressing their personal beliefs. I respect leaders such as Srikanth because to me, they were engaged at a very high level and had the confidence to show it.

The constitution of India does not specify a national language. Given the evolution of the country and the many ancient tribes involved, there are over 400 living languages across India. In some cases, people from adjoining states cannot understand each other's mother tongue. The Indian government officially recognizes 18 languages in India. However, Hindi and English are specified as their official languages.

One of the primary factors making outsourcing to India successful is the fact that we both share English as a common language. Even with a common language, it takes time for each side to learn how to listen and understand the linguistic peculiarities of the other. In the U.S. we tend to use sports analogies when we speak. Terms such as "full court press", or "keep your eye on the ball" that we use every day make absolutely no sense to many Indian nationals working here.

In Closing . . .

The Inside of Outsourcing will prove to be a valuable book for anyone embarking on an outsourcing project for the first time or looking for suggestions on how to improve an existing relationship.

I have enjoyed sharing my experiences and found the process of writing this book to be quite enjoyable as well. I sincerely believe that anyone after reading *The Inside of Outsourcing* will be left with a deeper understanding of the outsourcing process and how to decide if it is a viable alternative to address a current business problem.

Admittedly, outsourcing can be a contentious topic depending on which way the economic and political winds are blowing. *The Inside of Outsourcing* does not judge whether outsourcing itself is good or bad. The decision to outsource should be made on the merits of the business value inherent in doing so. The government should not, in any way, be involved in determining the intrinsic value of that decision. It was Henry David Thoreau that said "Government never furthered any enterprise but by the alacrity with which it got out of the way."

The Inside of Outsourcing took the reader, in an unbiased manner, on a journey through processes designed to help them develop their own conclusion by focusing on four primary topic areas.

➤ Building The Business Case: Outsourcing IT services should address a current business problem. Being able to define the problem was the first step covered in positioning an effective strategy. *The Inside of Outsourcing* helps the reader develop the Business Case, define the scope of the project and discuss how to sell the project to upper management. It also compared and contrasted the various sourcing models in use today. Finally, it demonstrated how to analyze costs and benefits by creating a detailed cash flow analysis of the project.

➤ Organizational Readiness: Organizations may need to make some internal changes in order to successfully implement and mange outsourcing projects. The readiness assessment included in *The Inside of Outsourcing* addressed the various aspects of outsourcing including IT Organization design; creating project objectives and how to manage staff impacted by the decision to outsource. Also discussed were the different managerial skills required as well as recommended structural changes such centralizing support and creating Supplier Relationship Management function.

➢ Supplier Selection: Identifying the right supplier and drafting the contract are important aspects of outsourcing especially with multi-year agreements. *The Inside of Outsourcing* helped the reader generate and manage the Request for Proposal in order to identify the best partner. Suggestions for what to include in the MSA and SoW's was discussed at length along with performance metrics used to manage the ongoing outsourcing agreement.

➢ Implementation and Governance: The final section of *The Inside of Outsourcing* described the implementation process. It described the roles and how to work through the various stages of Knowledge Transition. Also discussed was the linkage to the broader IT environment and supplier community that may be engrained in the organizational fabric. *The Inside of Outsourcing* thoroughly explains the roles and responsibilities of the Governance Team. This team must be in place to assure the client gets full value out of the contract while maintaining a positive relationship with the supplier. Finally, learning about the supplier's culture and contrasting it to the U.S. was, for me, one of the more enjoyable aspects of the outsourcing process.

If there are any questions about the content within *The Inside of Outsourcing*, or if you would be interested in a hosting a workshop on the topic of outsourcing IT services, please contact me via LaBella Consulting Inc. at:

LCIConsultingServices@gmail.com

www.ingramcontent.com/pod-product-compliance
Lightning Source LLC
Chambersburg PA
CBHW060005210326
41520CB00009B/833